Fundamentals
of
Entomology

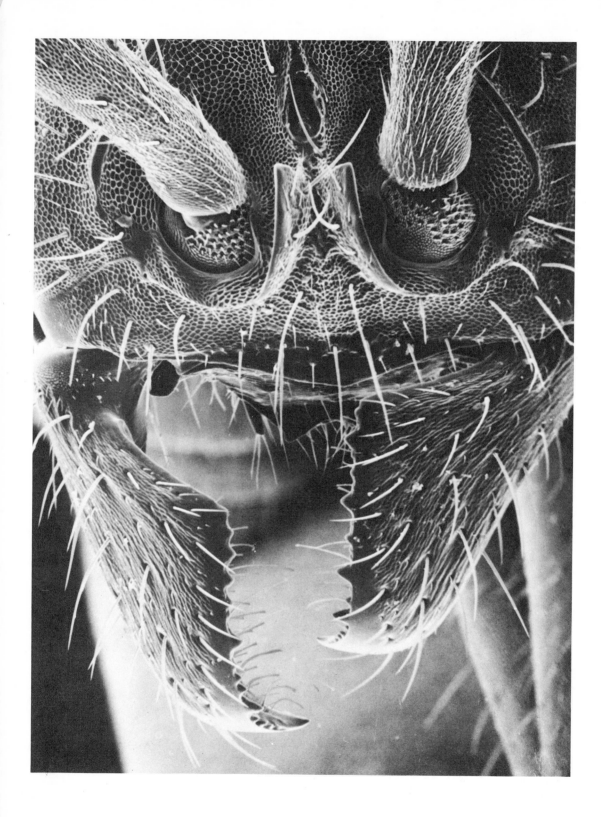

FUNDAMENTALS OF ENTOMOLOGY

Richard J. Elzinga

Department of Entomology
Kansas State University

PRENTICE-HALL, INC., Englewood Cliffs, New Jersey 07632

Library of Congress Cataloging in Publication Data

Elzinga, Richard J   (date)
    Fundamentals of entomology.

    Bibliography: p.
    Includes index.
    1. Entomology. I. Title.
QL463.E48   595.7   76-49874
ISBN 0-13-338186-2

Cover
Gall midge (Figure 118)

Frontispiece
The head of an army ant, *Eciton hamatum.*
(Photograph by author)

Printed in the United States of America

10   9   8   7   6   5   4   3   2   1

PRENTICE-HALL INTERNATIONAL, INC., *London*
PRENTICE-HALL OF AUSTRALIA PTY. LIMITED, *Sydney*
PRENTICE-HALL OF CANADA, LTD., *Toronto*
PRENTICE-HALL OF INDIA PRIVATE LIMITED, *New Delhi*
PRENTICE-HALL OF JAPAN, INC., *Tokyo*
PRENTICE-HALL OF SOUTHEAST ASIA PTE. LTD., *Singapore*
WHITEHALL BOOKS LIMITED, *Wellington, New Zealand*

Contents

v

Preface

Insects deserve study for many reasons. One of the major reasons is their unparalleled diversity: insects can provide a better understanding of nature and the many ways that biological problems have been met. Approximately 70 to 75 percent of the known species of animals are classified as insects, and 27 orders and about 600 families are found in North America north of Mexico (Borror and DeLong, 1970). Some insects live in the arid deserts, some in hot springs up to 80° C, others on mountain peaks as high as 6,096 m, some in tropical rain forests; and there are insects that live in arctic temperatures that reach below -20 °C. A second major reason is that a knowledge of insects is essential as we manipulate ecosystems for increased food production and better health. Back in the early 1900s, many entomologists were concerned about the competition for food between humankind and insects, and some entomologists believed that insect control was imperative for survival of the human race. Although such a position may seem somewhat extreme, insects do consume or spoil sufficient crops and products to feed the many millions of people who starve each year. And insect-transmitted diseases, both to humans and their crops, remain a threat to health and civilizations.

Five years ago students and faculty encouraged me to write an introductory text that condensed this diversity and influence of insects upon the ecosystem into a basic insect plan. The initial portion presents the fundamental structure-function of both external and internal structures. Upon this framework is superimposed the development and impact of the environment upon insects. Systematics is normally limited to the order and family level. Major interactions between insects, plants, and humans are presented, but control aspects are only discussed briefly because of the many detailed available texts on this subject and because of the current flux in the status of insecticide usage and other control methodology. A glossary is provided at the end of this book.

My gratitude is expressed to the many students, colleagues, and to my family, who have contributed to and encouraged this work. Acknowledgment is extended to those who loaned or gave permission to use illustrations and photographs. Scanning electron microscope photographs were taken in the SEM Laboratory, Kansas State University.

To those seeking knowledge of the basic insect strategy, I dedicate this text.

Richard J. Elzinga

Manhattan, Kansas

1

The Arthropod Plan

It has been said "To understand a person, one must also understand his family." Similarly, to understand insects (entomology) requires at least a cursory knowledge of other animals and especially those classified as the phylum Arthropoda (*arthros* = joint, *poda* = foot). Arthropods differ from many other invertebrates by having the following:

1. externally segmented bodies and appendages
2. appendages modified for feeding
3. an exoskeleton with chitin
4. a hemocoel instead of a coelom
5. no cilia
6. a ventral nerve cord and dorsal brain
7. bilateral symmetry

They are believed to have originated from annelidlike ancestors although transitional forms are lacking to substantiate this hypothesis. Three major lines of evolution seem to have occurred as indicated by the subdivision of the phylum into subphyla: Mandibulata, those that have well-developed *mandibles*; Chelicerata, those that utilize *chelicerae;* and Trilobitamorpha, known only from fossils, that apparently had none of their appendages specifically modified for feeding.

Arthropods are one of the most biologically successful groups of animals, for they live in the greatest variety of habitats, exhibit diverse types of locomotion, have the widest range of structural variations, eat the greatest variety of food, and include the greatest number of species.

SEGMENTATION AND TAGMOSIS

The ancestors of arthropods undoubtedly were bilaterally symmetrical and had their major sensory structures located at the anterior end of the body

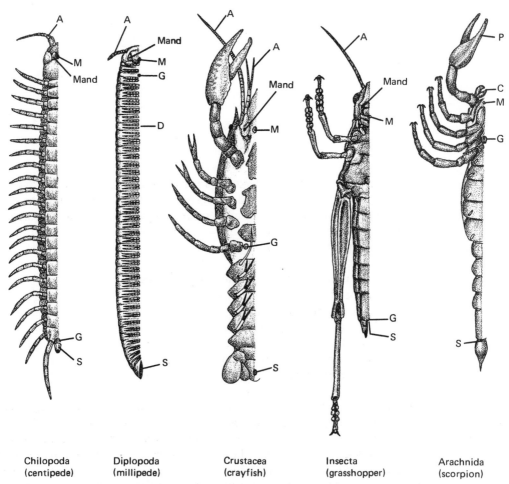

| Chilopoda | Diplopoda | Crustacea | Insecta | Arachnida |
| (centipede) | (millipede) | (crayfish) | (grasshopper) | (scorpion) |

Figure 1. Examples of the classes of arthropods and their body plan. A, antenna; C, chelicera; D, diplosegment; G, genital pore; M, mouth; Mand, mandible; P, pedipalp; S, anus.

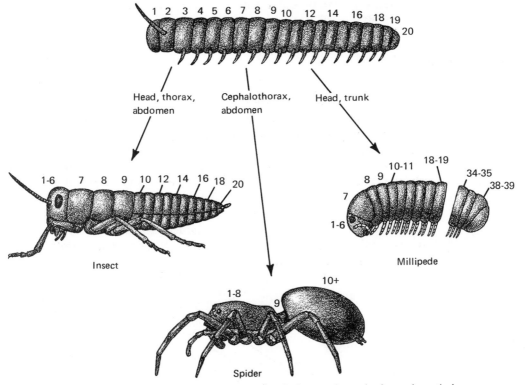

Figure 2. Segmentation and tagmosis of three arthropods from theoretical ancestor.

to perceive the forward environment. Their body probably consisted of either 20 or 21 *metameres* or segments, each of which possessed a pair of short lobelike appendages. The alimentary canal had two terminal openings, the anus and mouth.

From this archeotype evolved the many diverse shapes and forms of present-day arthropods, As seen in Figures 1 and 2, arthropod bodies are specialized into functional regions or *tagma,* a process termed *tagmosis*. In their primitive state, the anterior 6 segments evolved into the *head* (sensory, feeding, and coordination center), and the remaining segments or *trunk* retained their generalized role including locomotion (centipedes and millipedes). In more advanced states, the anterior 8 (spiders) to 14 segments (crayfish) were highly modified into a cephalothorax (sensory, feeding, coordination, and locomotor center), and the remaining segments became the abdomen and normally lost most of their appendages and role in movement. Another variation, found in insects and many crustaceans, resulted in three body regions, the *head* (sensory, feeding, and coordination center), the *thorax* (locomotion), and the *abdomen*. The localization of the locomotor area into the thorax or cephalothorax reduced undulation tendencies, such as those in centipedes, since the propulsion force of long legs is applied to a small area (Wells, 1968).

EXOSKELETON

One of the major requirements of animals is to slow down water uptake or loss from the body; this is often accomplished by the production of a slime or mucoprotein that covers the body surface. In arthropods, the body is covered by an exoskeleton or *integument.* Basically, the integument consists of a *basement membrane,* a layer of *epidermal cells,* and an externally secreted layer, the *cuticle,* which contains up to one-half the dry weight of an insect. One of the major characteristics of the arthropod cuticle is *chitin,* a substance closely related to cellulose. This nitrogenous polysaccharide has a tannish color and is flexible. The suit of armor so characteristic of this phylum results from the addition of certain material to chitin. If the protein *sclerotin* is added, as in insects or arachnids, the resulting hardening or tanning is termed

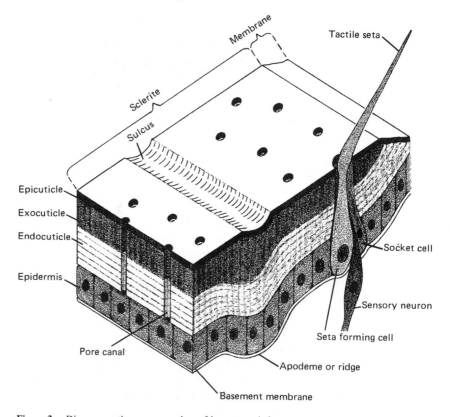

Figure 3. Diagrammatic representation of insect exoskeleton.

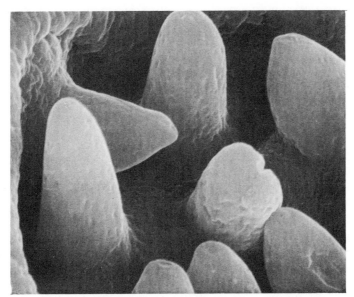

Figure 4. Chemoreceptor setae on maxillary palp of a cutworm caterpillar. The scanning electron micrograph permits magnification to the point of viewing the minute pores through which molecules can pass and contact the internal sensory cells.

sclerotization. This contrasts with the Crustacea and Diplopoda which add calcium, a process of *calcification.*

Cross sections through the cuticle (Fig. 3) reveal a laminate condition. The outermost multilayered *epicuticle,* consisting of lipids and polyphenols, provides waterproofing. Inside the thin epicuticle are the hardened *exocuticle,* the layer in which sclerotization occurs, and the flexible *endocuticle.* Areas that have a thick exocuticle are termed *sclerites,* and membranous regions consist primarily of endocuticle. *Pore canals* traverse the layers; their function is poorly understood, but they aid in the deposition of parts of the epicuticle.

Sensory receptors develop as part of the integument. Many appear hairlike and are termed *setae.* Although most setae are tactile receptors, e.g., movement indicates touch, some have been variously modified into *chemoreceptors* (respond to odors or taste) (Fig. 4). Other sensory structures include *tympanic* organs (hearing), temperature sensitive organs, and *photoreceptors* (light).

The benefits derived from the integument include protection from most chemicals except strong acids and bases, retardation of water movement both out of and into the body, high protection from physical damage and abrasion, a barrier to pathogens, a reservoir for some waste products, and an excellent structure for attaching a musculature system with good leverage. Disadvantages of the exoskeleton are that it necessitates special modifications for gaseous exchange, sensory pickup, and growth. Possessing an exoskeleton is a major restriction to growth, for only a limited amount of protoplasm can be added until the exoskeleton must be shed or *molted*. There are inherent dangers in molting, for the individual becomes vulnerable to physical and chemical forces as well as water loss during this period. The new exoskeleton must also be larger than the old one or the process would be self-defeating. The actual mechanics of these intriguing paradoxes will be discussed in Chapter 4.

SIZE

Humans seem to interest themselves in the grandiose things in nature. Dinosaurs, large snakes, and mammals are studied in preference to the average or minute forms. Similarly, no text would be complete without mentioning the large and bizarre examples of arthropods. As with many animals, these are usually found in the fossil record and are now extinct. The giant of the arthropods appears to be an aquatic arachnid, resembling modern scorpions, which attained the length of nearly 2 m. Another formidable fossil is the dragonfly with a wingspan of over 70 cm as illustrated in Figure 5. Contrast these with the more moderate yet still spectacular modern arthropods such as

Figure 5. Dragonfly reconstruction based on a fossil wing print from Permian fossils in Elmo, Kansas. The hind wing print is 7½ inches from the tip to base. Compare the reconstruction with a modern-sized dragonfly to the left.

the 32-cm wingspan in some moths, 33-cm walking sticks, 60-cm atlantic lobster, 27-cm centipede, and 30-cm millipede.

Most arthropods, however, were small in the past and remain so today. There are many advantages in being small since it is usually the small- to medium-sized species of most animal groups that have survived the long and rigorous geological history of our earth. The major advantages to being small include:

1. individuals require less energy and time to complete development
2. less energy is needed to sustain life both as individuals and as populations
3. it is easier to find protection from predators and other environmental extremes
4. numerous ecological niches are available for exploitation
5. muscular action is much more efficient and gravity has less of an effect
6. solar heat can be used to heat the body because of the high surface/volume ratios
7. the great ease of random dispersal by wind action

The main liability is the high potential water loss because of the high surface/volume ratio. Both the assets and liabilities of insect size will be explored in more detail in subsequent chapters.

SPECIATION

Arthropods have great adaptability and have radiated into most aquatic and terrestrial habitats. Trilobites, followed by the Crustacea, illustrate the dominant role of this phylum in the marine environment. Insecta and Arachnida have done the same in the terrestrial realm. Figure 6 summarizes

PHYLUM		ARTHROPOD CLASSES		MAJOR INSECT ORDERS	
*Arthropoda	842,000				
Mollusca	100,000			*Coleoptera	290,000
Chordata	45,000	*Insecta	715,000	Lepidoptera	114,000
Protozoa	30,000	*Arachnida	60,000	*Hymenoptera	113,000
Plathyhelminthes	15,000	Crustacea	50,000	*Diptera	86,000
Nematoda	10,000	Diplopoda	7,500	Homoptera	33,000
Coelenterata	9,600	Chilopoda	3,000	Hemiptera	25,000
Echinodermata	6,000	Misc.	6,500	Orthoptera	22,500
Porifera	4,200			Misc.	31,500
Ectoprocta	4,000				
Misc. Invertebrates	4,000				

Figure 6. Approximate numbers of animal species known. Some estimate the number of insect species as over 20 million, but most lists are more conservative. Asterisk (*) indicates areas where the greatest number of new species probably will be discovered.

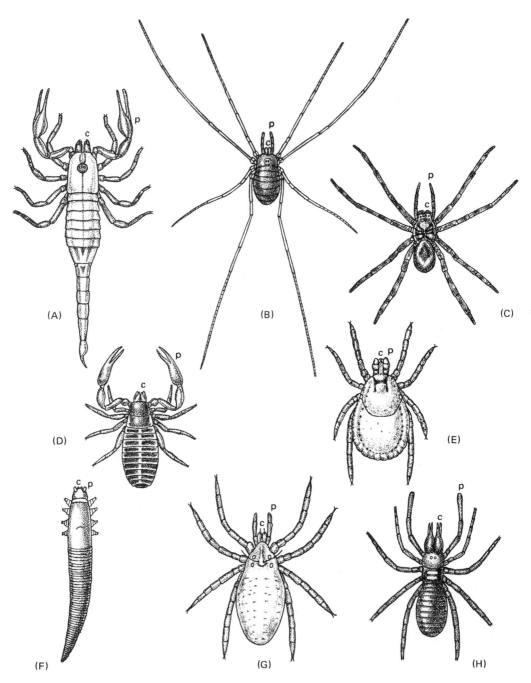

Figure 7. Diagrams of the more common arachnids. (A) scorpion; (B) harvestman; (C) spider; (D) pseudoscorpion; (E) tick; (F) follicle mite; (G) predaceous mite; (H) sunspider; c, chelicerae; p, pedipalps.

current knowledge of animal speciation and shows, to a great degree, the versatility of this group. Over 80 percent of the known animal species are arthropods, and an even higher figure is expected once all species have been classified. When the 800,000 are compared to the 350,000 plant species, the adaptive radiation becomes even more evident.

A cursory view of modern arthropods reveals five major groupings or *classes*. Species in each are more closely related to one another than to those in other groupings because of their tagmosis, origin of major feeding structures, modifications of appendages, location of genital openings, life cycles, and respiratory apparatus.

ARTHROPOD CLASSES

Arachnida

Included in this group of animals are the spiders, mites, ticks, scorpions, pseudoscorpions, harvestmen, whipscorpions, and sunspiders (Fig. 7). All are characterized by possessing *chelicerae* and *pedipalpi,* lacking antennae, having four pairs of legs and a cephalothorax and abdomen. Except for the mites and ticks, specialization only occurs for a carnivorous diet. Most are terrestrial.

Chelicerae (Fig. 8) are believed to be appendages of the third body segment, usually modified into predatory organs. Most have an opposable

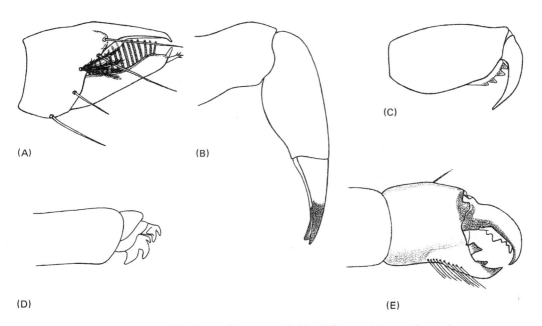

(A) (B) (C) (D) (E)

Figure 8. Modifications of some arachnid chelicerae. (A) pseudoscorpion; (B) harvestman; (C) spider; (D) tick; (E) scorpion.

Figure 9. Sunspiders or solpugids have greatly enlarged chelicerae and pedipalpi. The central chelicerae extend from the central eyes outward and are chelate, while the pedipalpi are leglike.

Figure 10. One use of silk by spiders is in covering eggs. This egg sac will moderate the influence of freezing temperatures, provide some protective concealment from predators, and hold the eggs together.

"thumb" (*chelate* condition) whereby the prey may be grasped and torn apart (Fig. 9); the exceptions are the many parasitic mites, ticks, and spiders in which the thumb is lacking (*unchelate* condition).

Pedipalpi are variously modified. In some, such as scorpions and pseudoscorpions, they are chelate and greatly enlarged for capturing the prey. In male spiders they become copulatory organs. In most, however, they are sensory structures and also form the base of a preoral cavity in which the chelicerae function.

Prey is captured in several ways. Some arachnids actively seek food, but others wait in ambush. The zenith of specialization occurs in web spinning spiders where sticky silk from abdominal spinnerets is spun into a snare for entangling unsuspecting invertebrates (Fig. 11). Regardless of the means of capture, much extraoral digestion of the prey occurs. Digestive fluids are pumped into the wounds (Fig. 12), mixed with tissues as the chelicerae tear them apart, and then sucked into the body by a muscular pump located in the cephalothorax.

Figure 11. Silk is often used in snare making in spiders. The circular orbs are sticky and trap insects flying or jumping into the web. The spider avoids the sticky webbing and moves about on the nonsticky spokes of the web.

Figure 12. Wolf spider (*Lycosa*) feeding upon cockroach. Large spiders such as this, which don't spin silken snares or webs, must capture prey by sheer force.

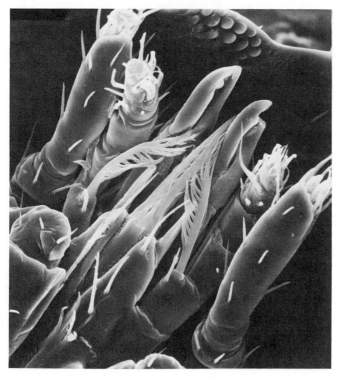

Figure 13. Scanning electron microscope photograph of the mouthparts of a mite (*Planodiscuts*) found on army ants.

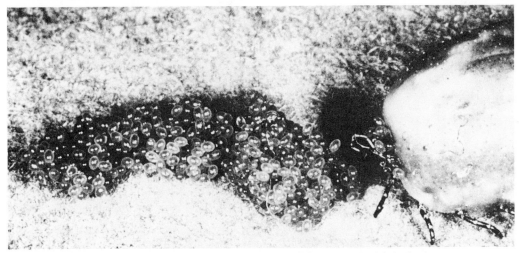

Figure 14. This fully engorged tick (*Dermacentor parumapertus*) is in the process of oviposition. Almost double the number of eggs already seen will be deposited.

Parasitic forms, ticks, and some mites, are highly specialized (Fig. 13). Although parasitic mites have been collected internally from the respiratory systems of both vertebrates and invertebrates, as well as from cloacas of birds and reptiles, most are located on the outer covering of the body. Hard-bodied ticks, for example, hatch in great numbers (Fig. 14) and attach themselves to a host by their mouthparts (Fig. 15). The "bite" of the chigger, the larval stage of a specific group of mites, is well known, but less obvious are the skin mites (*Demodex*) that inhabit pores in approximately 50 percent of unsuspecting Americans today. Most mites, however, are not parasitic but feed on plants and on other arthropods. Phytophagous mites penetrate plant tissues and use

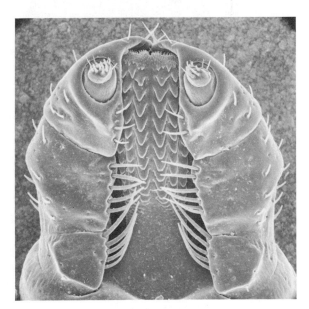

Figure 15. The toothed structure is the hypostome. Once this structure is in place and a "cement" secreted, the tick is difficult to remove.

Figure 16. The appearance of giant velvet red mites (*Dinothrombium* sp.) following a desert rain storm is a dramatic occurrence. A predator, this mite feeds on small arthropods, especially termites.

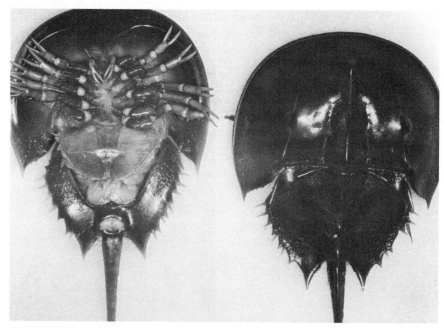

Figure 17. Some arthropods, particularly those that are aquatic and have physical support from water, reach large size as illustrated by this marine horseshoe crab (*Limulus polyphemus*).

their chelicerae to withdraw such liquids as sap, interstitial fluids, or lysed cells. The detrimental effect of these species upon our crops and shrubs by feeding and disease transmission is of economic concern. Less obvious are the mites that feed upon other mites, insects, and arthropod eggs and are beneficial to humans (Fig. 16).

Small arachnids obtain oxygen through their exoskeleton and/or minute tracheal tubes. The large species, however, have *book lungs,* internal sacs with many leaflike membranes, to increase the surface area for gaseous exchange and yet minimize water loss by maintaining this expansion internally. A few species are aquatic (Fig. 17) and utilize *gills.*

In most arachnids, sperm is transferred inside *spermatophores,* a sclerotized "suitcase." After fertilization, development takes two paths. Many, such as spiders and scorpions, hatch from the eggs resembling the adult and the successive molts produce only minor changes (Fig. 18). In others, such as the mites, more drastic changes take place as the young pass through successive larval (3 pairs of legs) and nymphal (4 pairs of legs) stages. The life cycle varies, but it is normally a function of size and temperature. Many mites expire after a few months, but a large tarantula may survive over 20 years in the tropics.

Bites and stings by arachnids are painful and are sometimes dangerous to humans. Spiders possess poison glands that empty through a duct into the terminal segment of the chelicera. These toxins normally kill prey; however, a few species of spiders, the black widow and brown spider species (Fig. 19), also contain enzymes that are harmful to humans and other vertebrates. Some scorpions also have toxins harmful to humans. The toxins are injected by the sting located at the terminal end of their abdomen.

Crustacea

This class consists of approximately 50,000 species, of which some are represented only as fossils. They are madibulate, i.e., possess as their major feeding structures appendages called *mandibles.* At least during some stage in their life history, their appendages are branched or *biramous.* The most anterior two pairs of appendages are modified into antennae that, in minute forms, are often used in swimming.

When viewing the class Crustacea, one is immediately amazed at the many diverse shapes and sizes (Fig. 20). In addition to shapes, tagmosis is extreme but is usually comprised toward either a head, thorax, and abdomen or a cephalothorax and abdomen. All harden their exoskeleton by the addition of calcium salts, hence the origin of the name Crustacea.

Although most Crustacea are marine, some inhabit freshwater (Fig.21) and a few are terrestrial. Because of their aquatic dependence, oxygen must be obtained from a medium in which it is in relatively low concentration. Small individuals, with their high surface/volume ratio, are able to take in oxygen solely through their body surface. Large forms have gills, expansions from

SPIDER Eggs in silken cocoon →	Hatch as larva (immobile) →	Molt and leave cocoon →	Normal habits as nymph or spiderling →	4–20+ molts → Adult
SCORPION Embryos in female →	Born as larva (ride on female) →	Molt and leave female →	Normal habits as nymph →	7–8 molts → Adult
HARVESTMAN Eggs in soil →	Hatch as nymph (nonfeeding) →	Molt →	Normal habits as nymph →	6–8 molts → Adult
PSEUDO-SCORPION Eggs in external female chamber →	Hatch as larva (attached to female) →	Molt and leave female →	Normal habits as nymph →	3 molts → Adult
MITE Eggs deposited variously →	Hatch as larva (nonfeeding, usually) →	Molt →	Normal habits as nymph →	1–3 molts → Adult
TICK Eggs deposited in soil →	Hatch as larva (feeding) →	Molt →	Normal habits as nymph →	1–8 molts → Adult

Figure 18. Representative life cycles of the more common Arachids.

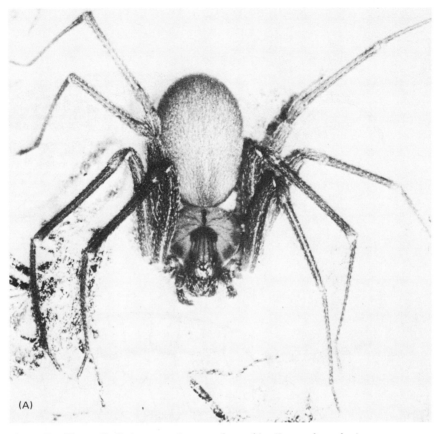

Figure 19. The medically important brown recluse spider (*Loxosceles reclusa*). (A) an overall view indicating the characteristic violin-shaped design on the dorsum of the cephalothorax; (B) posterior view of the chelicerae with the poison duct pore and semi-chelate modification.

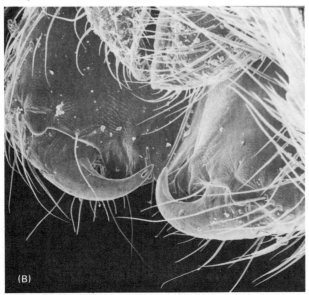

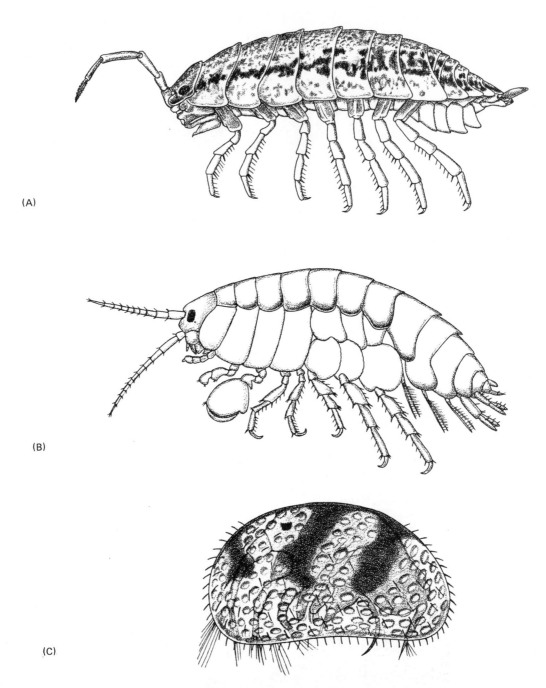

(A)

(B)

(C)

Figure 20. Variations within the crustacea. (A) terrestrial isopod; (B) aquatic sideswimmer; (C) aquatic ostracod.

Figure 21. This crayfish (*Orconectes* sp.) is contracting its abdomen, a means of rapidly propelling itself posteriorly away from danger.

their legs, which constantly circulate water across the increased surface area to aid in oxygen uptake.

The majority of crustaceans hatch from eggs in a shape very unlike the adult and pass through a series of larval stages. The simplest larva is the *nauplius,* consisting of an unsegmented body, three pairs of appendages, and a single median eye. Species that initiate life with this stage usually progress to the adult through a series of progressive steps separated by molts. At each molt, segments and appendages are added at a "budding zone" near the posterior end of the body until the predetermined adult number is reached. Such a process is termed *anamorphic development.* The three swimming appendages of the original nauplius eventually become the two pairs of antennae and the mandibles of the adult.

Some of the more advanced Crustacea, however, hatch as specialized larvae and by pass the nauplius through comparatively large changes at each molt. As an example, crabs have a *zoea* larva in which many of the abdominal segments are already present at hatching and the mouthparts are functional.

Food often varies according to the size of the individual. Larval stages and small adults are usually dependent on other plankton such as diatoms. The larger crustaceans are usually carnivores or scavengers, although a few are parasites.

Chilopoda

Centipedes have a long body divided into a *head* and a multisegmented *trunk.* The head has a pair of moderately long antennae, a pair of mandibles,

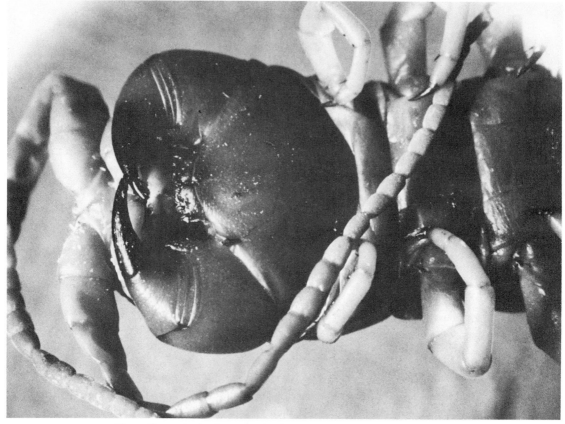

Figure 22. A centipede head and poison claw.

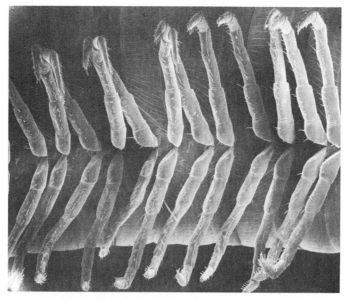

Figure 23. A scanning electron microscope photograph revealing the diplosegments of a millipede.

and 2 pairs of maxillae. With the exception of 2 terminal segments, the remaining segments, varying up to 190 in number, each have a single pair of legs. Some species have short legs that are very efficient in transporting the body through narrow burrows and cracks as they chase prey, but other species have elongated appendages for running in more open situations. The first pair of legs behind the head is modified into poison claws (Fig. 22) and is used in capturing animals upon which they feed.

Eggs are deposited either singly or in groups; if they are oviposited in groups, the female will usually guard the eggs. Primitive species hatch with few segments and add segments during successive molts, a process mentioned previously as anamorphic development. The maximal number is reached at 15 pairs of legs.

The more advanced centipedes are born with the same number of segments as the adult, a condition termed *epimorphic development.* Curiously, these include the species with the greatest number of segments and appendages. Molting in these organisms results only in a size increase with the maximal size of approximately 275 mm recorded for one species in South America.

Centipedes usually hide during the daytime but are active at night. Their food consists of other arthropods, earthworms, and slugs. Although efficient in killing their prey, the toxin from the poison claws is rarely dangerous to humans.

Diplopoda

The millipedes represent another group that have their body differentiated into a head and trunk. The head contains a single pair of short antennae, a pair of mandibles, and a single pair of fused maxillae, the *gnathochilarium.* Unlike the centipede, most of the trunk segments are actually *diplosegments* (Fig. 23) that arose from the fusion of two segments. The genital pore is located anteriorly.

Eggs are deposited in groups of from 25 to 250 in soil or moist humus. Larvae possess only 3 pairs of legs at hatching, but diplosegments are added by anamorphosis in groups of up to 3 or more at a molt until the predetermined number is reached. A maximal disposegment number of 40 and body length of 300 mm is recorded. The life cycle is completed in from 1 to 7 years.

Molting often occurs in earthen chambers (Fig. 24) in which water loss is minimized and in which protection from predators is maximized. The exoskeleton, unlike that of the centipedes but like that of the crustaceans, is hardened by incorporating calcium carbonate.

Most millipedes are found in rotting logs, leaf litter, humus, and under stones where they feed mainly on decaying vegetative material and fungus. Since the legs are numerous, close together, and short, locomotion is relatively slow and occurs by waves of leg movement down the trunk in synchronous fashion. When disturbed, they often roll into tight coils to gain maximal pro-

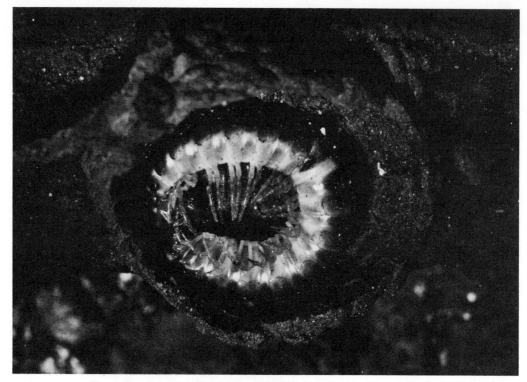

Figure 24. Earthen cell in which millipede is estivating during Costa Rican dry season. Such behavior reduces water loss to a minimum.

tection from their hard exoskeleton. Many secrete noxious fluids, such as hydrogen cyanide, from specialized glands to repel predators.

Insecta

These are arthropods that have 3 body regions, 3 pairs of thoracic legs, and a pair of antennae, and that feed by mandibles. The number of body segments consists of from 19 to 20 in most species. They are the only invertebrate with wings. These wings have had a marked influence upon their success. Most insects are terrestrial, although a few occupy aquatic habitats during part of their existence, and only the ocean has eluded extensive exploitation.

Development is epimorphic, except in the order Protura, and no new segments are added after hatching from the egg. Changes vary from minor (incomplete metamorphosis) to drastic (complete metamorphosis). Most insects complete development in a year or less but a few require up to 17 years to mature.

Special attention to the great number of insectan species should be made. Why have they become so successful? All the reasons are obviously not known but some of the major ones are probably as follows:

1. their small size
2. modification and exploitation of appendages into the many types of mouthparts and locomotion–food gathering legs
3. extensive development of complete metamorphosis in which the immatures and adults have evolved to feed on different foods and hence are not in competition
4. rapid life cycles wherein mutations can be rapidly selected for and incorporated into the population gene pool
5. the many different species-isolating mechanisms involving genital, hormonal, and behavioral modifications.

The success story of this group is one that all biologists should be exposed to and it is one that will be unfolded in the subsequent chapters.

Questions

1. What is an arthropod? How does it differ from other invertebrates?
2. What are the advantages and disadvantages of an exoskeleton? How are the disadvantages overcome by arthropods?
3. What is tagmosis? Of what advantage is this process? What are the evolutionary trends as seen in Crustacea, Insecta, and Arachnida?
4. How does the number of species represent evolutionary success?
5. Is small size advantageous? What are the selective pressures that interplay and affect size in arthropods?
6. Which are the most closely related, spiders and scorpions or centipedes and millipedes? What evidences are available that you can use to determine which are most closely related? How are relationships determined?
7. Compare the classes of arthropods from a morphological, developmental, and habitat preference viewpoint.
8. What is the difference between anamorphic development and epimorphic development? In which class is each found?
9. Are any arthropods of potential danger to humans? In what way? How does such a knowledge affect your life?

2

The Insect Externally

THE HEAD

The most anterior region or tagma of the insect body is the head. Embryologists have noted that heads are formed differently in various animal groups, but, in each instance, there is an assemblage to carry out (1) ingestion of food, (2) major sensory perception, (3) coordination of bodily activities, and (4) protection of the coordinating centers. The sensory structures are located near the mouthparts and coordinating centers for more effective food ingestion and to enable more rapid responses to incoming stimuli. The head, therefore, is defined as a *functional unit* responsible for carrying out the above four processes.

In insects the individual segments of the head have fused so completely that external evidence of their presence or origin is lost. How then can we determine its segmentation? As with most animals, primary evidence of origin comes from studying embryology and the nervous system. Research has found that these two factors are the least likely to be modified by adaptive selection, but we expect and find in as diversified a group as insects that even these have undergone sufficient change to prevent decisive conclusions. However, without going into supportive data as found in Matsuda (1965), most entomologists currently believe that the insect head consists of the first six metameres or segments; this assumption will serve as the basis for further discussions.

Protection and Feeding Role

As in most animals, the insect head is covered by hardened elements to assist both in protection of the coordinating centers and in the process of feeding. As expected, the least modifications occur in primitive or the least specialized species in which linear invaginations or *sulci* (sometimes referred to as *sutures*) may be seen along the base near the mouthparts and extend dorsally. These folds are braces to prevent collapse of the head during feeding or when subjected to external physical stresses. If one creases paper or any flattened yet semirigid material, increased strength is gained. Similarly, sulci can brace heads without substantial thickening. Since evolutionary selection, however, has resulted in harder and thicker head skeletons in the advanced insects, the importance of sulci has diminished accordingly.

The exoskeleton between the sulci are termed *sclerites,* which have no segmental value and are useless in elucidating head segmentation. As seen in Figure 25, the sclerite located in the "forehead" region is termed the *frons*. Below this plate is the *clypeus* and the flaplike *labrum*. Beneath the compound

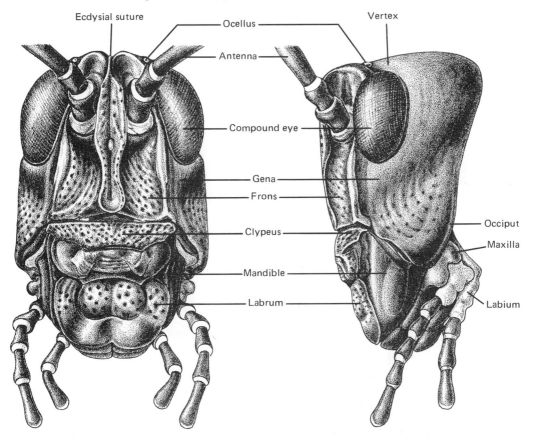

Figure 25. The insect head illustrating the major sclerites, mouthparts, and sensory structures.

eye is the *gena* and above is the *vertex*. A sulcus separates the gena from the posterior *occiput*. Only the ventral (for feeding) and posterior (for communication with the remaining portion of the body) parts of the head capsule remain uncovered by sclerites.

Are all external linear lines on the cranium sulci? In immatures, there exists between the antennae a Y-shaped *ecdysial suture* that extends posteriorly to the membranous neck or *cervix*. This suture has no exocuticle and hence cannot serve as a brace. It fractures during molting (Fig. 26) and permits the old head capsule to be shed.

Bracing of the head occurs internally by means of the *tentorium*. This structure is formed by invaginations of certain sulci at the base of the head capsule and, because of its shape, has great strength and assists in preventing collapse of the head.

As indicated previously, a progressive loss of sulci occurs with the concurrent thickening of the head capsule and evolution of the various mouthparts. The thickening or increased sclerotization now takes the stresses once imposed on the sulci, and the once obvious sclerites become poorly delineated regions as the sulci become indistinct. Also, the head position may become modified from the early *hypognathous* condition, in which the mouthparts are at right angles to the body axis, to the more advanced *opisthognathous* condition in which they project backward between the legs. Another modification, the *prognathous* condition, is commonly seen in carnivorous and/or forms that burrow in wood or soil. When the insect is prog-

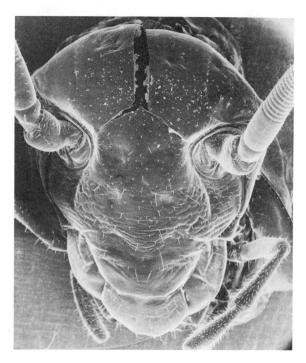

Figure 26. Scanning electron microscope photograph of a cockroach molting. Note the split along the ecdysial suture, which will permit the newly formed cranium to escape.

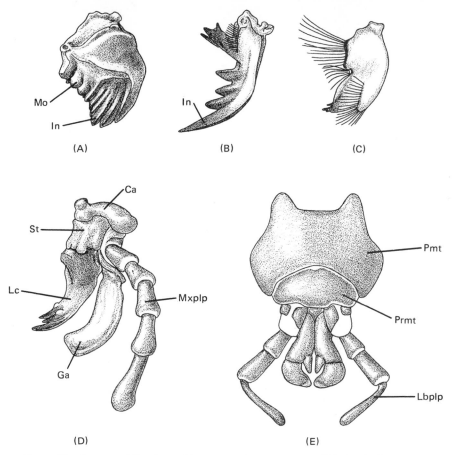

Figure 27. Some modifications of insect mouthparts. (A) mandible for feeding on plant material; (B) mandible of carnivorous insect; (C) mandible of scavenger modified for filtering food from water; (D) maxilla of herbivore; (E) labium of herbivore; Ca, cardo; Ga, galea; In, incisor; Lc, lacinia; Lbplp, labial palp; Mo, mola; Mxplp, maxillary palp; Pmt, prementum; Prmt, postmentum. (A, D, and E redrawn with slight modifications from Snodgrass, 1935.)

nathous the prey can be grasped easily in cramped quarters with little danger to the predator, or the burrow can be easily extended because of the projecting mouthparts.

Feeding in multicellular organisms often requires abrasion or chewing of the food. In mollusks this is accomplished by means of a tonguelike radula, in chordates by jaws and teeth, but in arthropods by appendages (Chap. 1). The primary parts in insects are the *mandibles* (Fig. 27A, B, C) or paired appendages of the fourth segment. Supporting their action are the *maxillae* (Fig. 27 D) or paired appendages from the fifth segment and the *labium* (Fig. 27E) or fused appendages of the sixth segment. Other parts may be involved and include the *labrum* and *hypopharynx* (Matsuda, 1965).

Are all insectan mouthparts similar? This can be answered by asking a second question, "Are all diets of insects similar?" One of the evolutionary marvels has been the numerous modifications that have permitted these invertebrates to feed upon nearly all organic materials. For simplicity, let us view some of these specializations under six major categories: chewing, cutting-sponging, sponging (Fig. 28), siphoning, piercing–sucking, and chewing-lapping, remembering that there are many variations of each. Note that only the chewing type is specialized for uptake of solid materials although many of the others may require penetration of solids to contact the liquid food.

Chewing type.   These are the least specialized. Let's use the June beetle larva (Fig. 29) as an example and view both structure and the feeding process together. The distal portions of the heavily sclerotized mandibles have cutting edges, and their bases have grinding surfaces. Note the arrangement of distal cutting surfaces and note that proximal grinding is similar to the tooth position and function in the jaws of mammals. The mandibles move across one another scissorlike. The muscles operating the mandibles are the largest in the body and normally originate from the vertex. Because of their chewing action, the head must either be braced by sulci or have thickened sclerotization.

    The anterior labrum, the mandibles and maxillae at the sides, and the posterior labium form a *preoral cavity* between the mouthparts in which chewing occurs (Fig. 30). Food is pulled into this cavity by the maxillae and cutting action of the mandibles. Saliva is secreted between the labium and hypopharynx (Fig. 30) and mixed with the food during chewing to assist in swallowing. The food bolus is then forced by the maxillae and hypopharynx up into the mouth for swallowing.

    Are all mouthparts the same in the chewing type? There are many dif-

Figure 28. Scanning electron microscope photograph of the labellum or "sponge" portion of the face fly proboscis. Note the canals or pseudotracheae for channeling fluids.

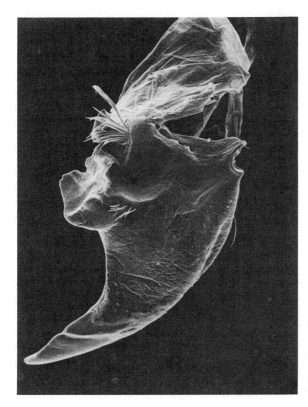

Figure 29. The mandible of a June beetle larva as seen through the scanning electron microscope. Note the enlarged molar area for extensive grinding.

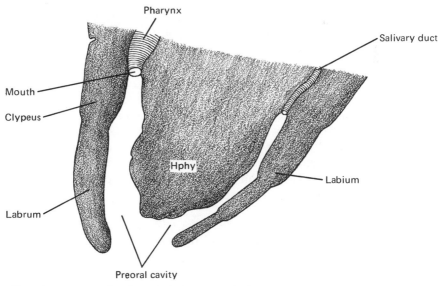

Figure 30. Cross section through the insect head indicating the preoral or chewing cavity and openings into the alimentary canal and salivary duct. (Redrawn with slight modifications from Snodgrass, 1935.)

ferences, but a few generalizations may be made. Predaceous insects usually have mandibles that are long, pointed, and very sharp (Fig. 27B) These are used, similarly to canine teeth in mammals, in killing the prey. Since little grinding is required, the molar area is noticeably reduced or absent. Just as mammal herbivores have different tooth modifications resulting from eating different foods of various consistencies, so also do insectan herbivores; those that feed on grasses or seeds have greater grinding surfaces on the mandibles than do those that chew soft succulent leaves.

In all instances, feeding is actually the culmination of a series of behavioral responses to stimuli. Using a phytophagous insect as an example, we see that the sequence typically follows the pattern of first being attracted to the food by odor, sensing the plant through contact receptors on the tarsi, taking a test bite whereby the taste receptors, especially those on the palpi (Fig. 4), can discriminate, and then feeding.

Cutting–sponging. This type (Fig. 31A) is restricted to a limited number of adult flies feeding as parasites upon blood from mammal hosts. Black flies and horseflies are good examples. The mandibles and maxillae are elongated, pointed, and function as *stylets* to pierce the skin. Once the capillary networks are pierced and disrupted, blood is released, sponged up into the labium with its many pseudotracheae, and sucked into the body through a specialized food canal between the labrum and hypopharynx. Anticoagulants, found in saliva, are pumped into the wound to prevent blood clotting and may keep the blood flowing from the wound for some time subsequent to the departure of the fly.

Sponging. Only adults of the more specialized flies, such as the house fly, have this highly modified type (Fig. 31B). The proboscis consists primarily of the labium; the mandibles and maxillae, except for the maxillary palpi, are either lost or incorporated into the basal elements. During feeding, the proboscis is lowered and salivary secretions are pumped out onto the food. The dissolved or suspended food is then moved by capillarity through the numerous pseudotracheae (Fig. 28) to a median reservoir in the "sponge" or *labellum.* The sucking pump now draws up the mixture through the labral–hypopharyngeal tube into the alimentary canal.

Tasting is carried out by both sensory hairs on the feet or tarsi and on the labellum. The scanning electron microscope photograph (Fig. 4) has recorded receptors similar to these.

Siphoning. Almost every naturalist has observed a butterfly or moth land upon a flower, uncoil, and straighten its proboscis. If nectar is present, this fluid is sucked into the body (Fig. 31C). The proboscis then coils up because of its natural elasticity and the next flower is visited.

The proboscis is highly specialized, consisting of only the greatly elongated galea of the maxillae held together medially by interlocking spines and hooks. Traversing this tube is the food canal and salivary duct.

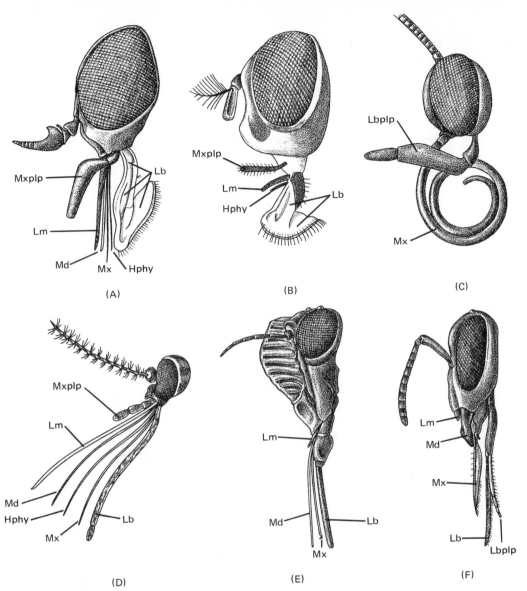

Figure 31. Some modifications of insect mouthparts from the basic chewing type. (A) cutting-sponging of a horsefly; (B) sponging of the house fly; (C) siphoning of a butterfly; (D) piercing-sucking of a mosquito; (E) piercing-sucking of a cicada; (F) chewing-lapping of the honeybee; Hyphy, hypopharynx; Lb, labium; lbplp, labial palp; Lm, labrum; Md, mandible; Mx, maxilla; Mxplp, maxillary palp.

Piercing–sucking. A wide variety of distantly related insects have this type (Fig. 31D, E). Herbivores such as the cicada, parasites such as fleas and mosquitoes, and carnivores such as assassin bugs are but a few examples of modifications for different diets. The basic plan is similar, however, and consists of a series of stylets (mandibles, maxillae, and sometimes the hypopharynx and labrum) enclosed by the sheath or *labium* to hold the stylets in position. The number of stylets varies from three in sucking lice to six in the mosquito. These move up and down piercing tissues. Molecular cohesion of the water molecules surrounding the stylets aid in holding the stylets together, although they are dovetailed by longitudinal ridges and grooves in some species in which up and down movements are permitted while preventing the stylets from parting. As the stylets penetrate deeper and deeper, the labium is often bent elbow-shaped out of the way and does not enter the wound (Fig. 32).

After penetration has been effected, a salivary secretion is injected from a *salivary syringe* and functions as a toxin in carnivorous species, functions as an anticoagulant in many parasites, or has a necrotic action on tissues in many insects that feed on plants. Released host fluids are sucked up through a food canal either in one of the stylets or through a groove formed between two of the stylets. Diseases are commonly transferred by insects possessing these modifications.

Figure 32. *Aedes aegypti* (L.) probing into the tissue of a frog's foot at the left and penetrating a capillary to ingest blood at the right. (Redrawn from Gordon and Lumsden, *Ann. Trop. Med. Parasit.,* 33:259-278. By permission of the Liverpool School of Tropical Medicine.)

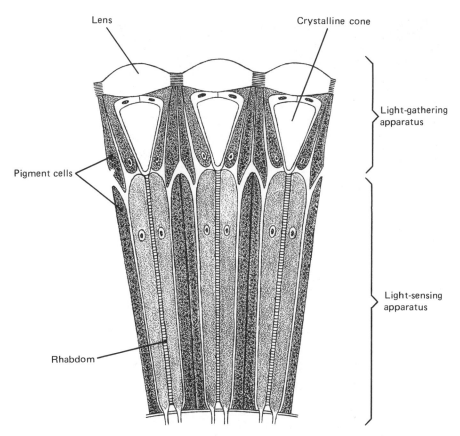

Figure 33. Diagrammatic representation of three ommatidia of the compound eye. (Redrawn with slight modifications from Snodgrass, 1935.)

Chewing-lapping. Adult honeybees (Fig. 31F) and bumblebees have mouthparts that are modified in still another fashion in order to utilize liquid food, in this case nectar and honey. The major feeding apparatus consists of a maxillo–labial complex. Surrounding the central "tongue," the glossae of the labium, is a tube formed from the galeal part of the maxillae. With the combined action of both the sucking pump and "tongue" moving up and down, nectar is drawn up into the body. The mandibles usually do not function directly in feeding but may be used not only for cutting flowers that have long corolla to gain access to the nectar but also for defense and molding wax into combs for storing honey in the hive.

Sensory Role

Compound Eyes. As with most animals, photoreceptors are usually located near the mouth at the anterior region of the body. The most conspicuous of these photoreceptors are the *compound eyes* of adults and many immatures called nymphs (Fig. 85). Each photoreceptor consists of a number of separate receptors or *ommatidia* (Figs. 33, 34); the number varies from a single om-

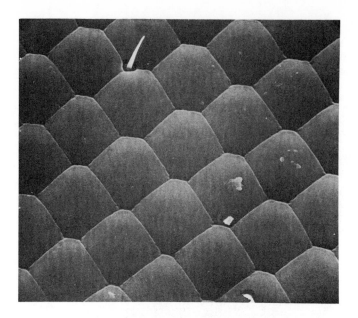

Figure 34. Scanning electron microscope photograph of the compound eye of a wasp (*Vespula maculata*) showing the individual ommatidial lenses.

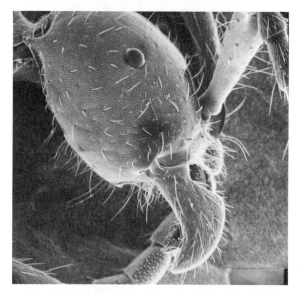

Figure 35. The head of an army ant (*Eciton* sp.) as seen through the scanning microscope. Only a single ommatidium is present indicating an essentially blind situation as far as image formation is concerned.

matidium in some ants (Fig. 35) to over 30,000 in some dragonflies. In most day-flying insects, each ommatidium has a light-gathering apparatus, the corneal *lens* and *crystalline cone,* and a light-sensing apparatus, the *rhabdom.* Direct light is focused onto the rhabdom, which contains visual pigments, and initiates a discharge of a nerve. Interpretation of these messages by association centers in the brain is made and the insect accomplishes vision.

In general, predators and fast-flying species that seek flowers or mates during flight have the greatest number of ommatidia, but soil inhabitants and occasional fliers tend to have the least. One reason for such an increase seems to be the advantage of increased depth perception. If an insect faces an object, the intervening distance is calculated by the angle between the eyes. As the two come closer, the ommatidia nearer and nearer the meson are used, thereby providing a measure of distance. The greater the number of ommatidia, the more precise will be this measure. Objects laterally positioned cannot be seen by both eyes and hence distance cannot be adequately measured.

Complexity and variation of the compound eye cannot be overstated. Variations from the above simplified explanation include:

1. insects active at night have ommatidia different from those that fly during day hours
2. eyes of fast-flying insects differ from those flying slow
3. eyes of different insects vary in their capability to perceive different wavelengths, although most are able to see ultraviolet and few can distinguish red wavelengths.

Other Photoreceptors.   Located on the frons or vertex of many adults and nymphs are the *ocelli*. A maximal number of three per individual are known. Ocelli probably do not perceive form because light is not focused on the sense cells. Their function is believed to be one of increasing body tonus, preparation to jump or fly, in response to drastic changes of illumination, such as would be produced by a shadow of an approaching predator. In addition, daily rhythms are affected when they are experimentally covered with paint or light reception is interfered with in other ways.

Larval insects, such as caterpillars, have photoreceptors structurally intermediate between typical ocelli and ommatidia. These *stemmata* (Fig. 36) are

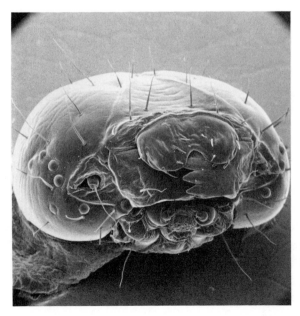

Figure 36. A scanning electron microscope photograph of a noctuid caterpillar head illustrating the relative position and number of stemmata.

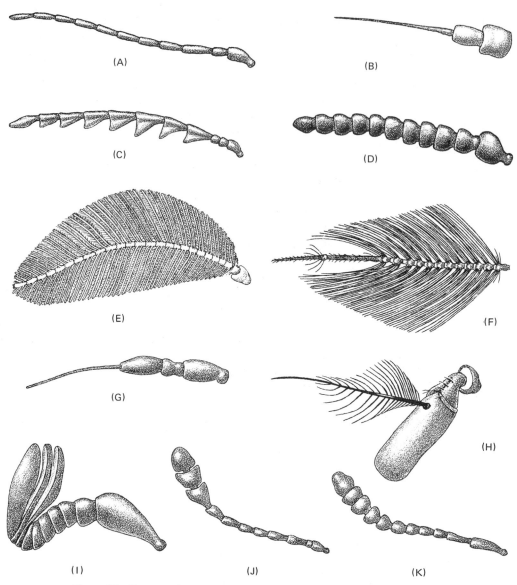

Figure 37. Types of insect antennae. (A) filiform; (B) setaceous; (C) serrate; (D) moniliform; (E) pectinate; (F) plumose; (G) stylate; (H) aristate; (I) lamellate; (J) capitate; (K) clavate.

usually in 2 scattered groups on the gena. They probably have weak powers of form perception because light can be focused, but visual acuity is probably low since there are usually fewer than 12 stemmata. During metamorphosis to adulthood, stemmata are lost and are replaced by the adult compound eye.

Antennae. Few people have picked up an insect without noticing the conspicuous "feelers" projecting from the head. Although reduced in many immature forms, these *antennae* (Fig. 37) are frequently large in adults in order to aid in the increased sensory activities necessary for the specialized food and mate location at this stage in the life cycle.

Only the two basal segments are operated by direct musculature. The remaining segments, often termed the *flagellum,* maintain their position by blood pressure and normal rigidity and move only when environmental factors contact and move them physically. The animal, then, interprets this contact and reacts accordingly.

In addition, more delicate tactile, odor, humidity, hearing, and various other stimuli are detected by appropriate receptors found on these appendages. Tactile setae are solid and have a sense cell at the base to detect movement. Other sensory setae are hollow and have sensory cells extending to openings for chemoreception (Fig. 4). Some sensory structures are not located in setae but are beneath the exoskeleton for determining pressure.

Many shapes of antennae may be seen with some of the major types or variations illustrated in Figures 37, 38, 39. Some are undoubtedly modified for increasing surface areas for receptors, but many plumose antennae have no more receptors than filiform types. Future research in this area will undoubtedly give us insight into correlating shape with function.

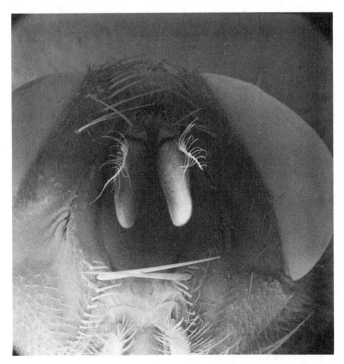

Figure 38. Antennae of a face fly (*Musca autumnalis*). Located to the side of the antennae is the ptilinal groove through which the ptilinum was everted to push off the cap of the puparium (see Figure 91).

Figure 39. The terminal segment of an army ant (*Eciton* sp.) antenna. Note the disproportionate position of the chemoreceptors (peglike and many to the right). Since the ant is essentially blind and utilizes its antennae to follow chemical trails, the majority of the sensory setae are located ventrally where they constantly touch the substrate as the insect moves about.

THE THORAX

The function of locomotion in adult and nymphal insects has been taken over almost exclusively by three body segments, collectively referred to as the *thorax* (Fig. 40). For ease in discussion, the seventh body segment will be referred to as the *prothorax,* the eighth as the *mesothorax,* and the ninth as the *metathorax.* Each segment is usually sclerotized to maintain a more or less rigid position and prevent the body wall from flexing during movement of the appendages. The basic plan of each of the thoracic segments consists of a dorsal *tergum* or notum, a ventral *sternum,* and two lateral *pleura* (pleuron = singular). The pleura are the result of incorporating a leg segment into the body wall and each is braced by a linear invaginated ridge, the *pleural sulcus.* The basal part of the sulcus serves as a point of articulation for the legs and braces the pleural plates against the tension of muscles which originate on the tergum and move the legs. For descriptive purposes, the pleural plate anterior to the sulcus is called the *episternum,* the posterior one the *epimeron.* A pair of *spiracles,* openings into the respiratory system, are found between the prothorax and mesothorax and between the mesothorax and the metathorax.

In addition to movement by appendages, most advanced adult insects are also able to fly. Two of the thoracic segments, the mesothorax and metathorax, have wings that provide a means of locomotion that has carried insects into nearly all possible habitats. Wings arose not as modified appendages, as in other animals capable of flying, but from outgrowths of the tergum. Because the tergum or notum is involved in flight (see Chap. 3), it has become highly modified in winged species. The other sclerites of the mesothoracic and metathoracic segments are also altered but to a lesser degree. Both the pleura and sterna tend to fuse solidly with one another and between

38

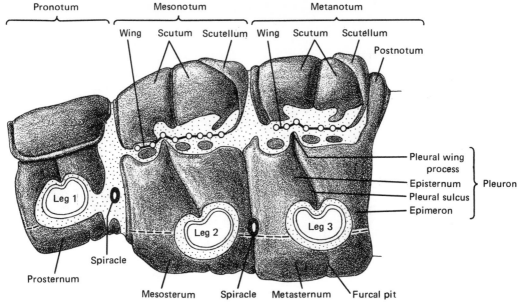

Figure 40. Diagrammatic representation of the insect thorax indicating the three segments involved as well as sclerites and position of legs and wings.

the wing-bearing segments to form a U-shaped brace beneath the wings. The dorsal extension of each pleural sulcus, the *pleural wing process,* functions as a fulcrum for wing movement. Often the sternum has been invaginated into the body to form a furca for more efficient bracing and muscle attachment as well as to enable the legs to be placed directly under the body for better movement.

Wings

Wings are first evident as *wing pads* in immature insects. In wing pads that develop externally, a lateral evagination occurs which has both upper and lower cuticula and epidermis (Fig. 41). Between these two layers may be seen tracheae, nerves, and an extension of the body cavity. With each successive molt, the pads enlarge until they appear as miniature wings (Fig. 42). After the last molt, blood pressure enlarges these structures, and fully developed wings of the adult individual result. During expansion, the epidermis is stretched and disorganizes, thereby allowing the two cuticular layers to fuse and sclerotize around the tracheae and nerves to form longitudinal *veins* and *crossveins* which serve as braces for the wings. Veins are valuable in demonstrating relationships and are so named to permit detailed studies on classification (Fig. 43).

Fossil records give little information about the evolution of wings. There are, however, some fossil insects with short, winglike lateral expansions of the tergum, the *paranota,* and insect wings are believed to have evolved from these projections. Most evidence on strategies comes from studying modern wings

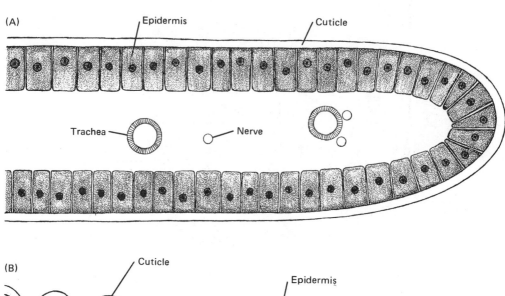

(A)

Epidermis

Cuticle

Trachea

Nerve

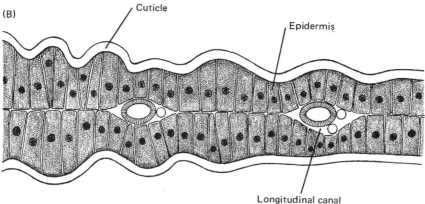

(B)

Cuticle

Epidermis

Longitudinal canal

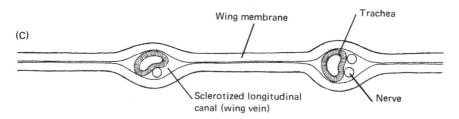

Wing membrane

Trachea

(C)

Sclerotized longitudinal
canal (wing vein)

Nerve

Figure 41. Development of the insect wing as seen in cross section. (A) during
the early pupa; (B) during the late pupa; (C) during the adult.

and hypothesizing the most likely avenues of evolutionary change (Fig. 44).
Wings are not solidly attached to the body as in airplanes or in paranota. In
the more primitive insects, wing movement is restricted to essentially one
plane; dragonflies serve as an example of this *paleopterous* condition. In most

40

Figure 42. The immature long-horned grasshopper nymph possesses developing wings in the form of wing pads. Note the shape and size. During subsequent molts they will enlarge until they form functional wings after the final shedding of the exoskeleton.

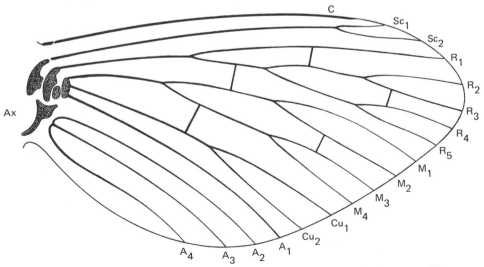

Figure 43. Generalized wing illustrating venetion. A, anal veins; Ax, axillary sclerites; C, costa vein; Cu, cubitus veins; M, median vein; R, radius veins; Sc, subcosta veins.

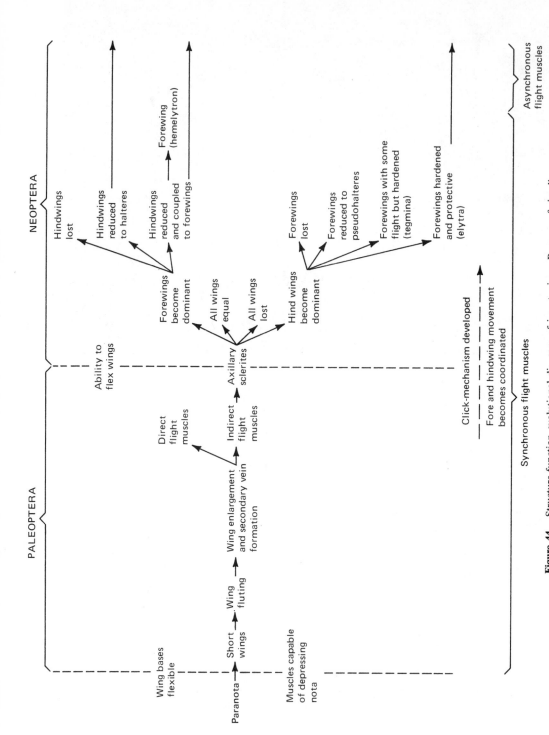

Figure 44. Structure-function evolutional diagram of insect wings. Because of simplification, phylogeny is not necessarily correlated.

insects, however, wings are capable of being *flexed* or folded posteriorly over the body when not in use and hence can be moved on more than one plane. The flexing mechanism results from the development of *axillary sclerites* in the membrane at the wing base. These *neopterous* wings give an insect the advantages of flight while also permitting these structures to be folded to enable the individual to crawl under bark or rocks and to hide from predators or extremes in weather without damaging the wings.

Early in wing evolution, wings had extensive *fluting* or longitudinal creases with many veins for support (Edmunds, 1954). The rigidity obtained from fluting permitted flight, but sculling action, a common figure-8 movement of the wings, is interfered with and drag results. Today most highly developed wings, as seen in flies, are slender with a few strong anterior veins for strength and permitting sculling.

Usually one pair of wings become dominant. In some insects the hind wings carry much of the burden of flight while the forewings become increasingly sclerotized in order to protect the wings when they are flexed. Forewings so modified are called *tegmina* (grasshoppers) when they are leathery in texture (Fig. 45) or *elytra* (beetles) when they are heavily sclerotized.

Figure 45. Wings of a grasshopper (*Dissosteira carolina*). The forewing or tegmina is mainly protective during periods of rest. The hind or flight wings are large and fluted to give strength to the large surface.

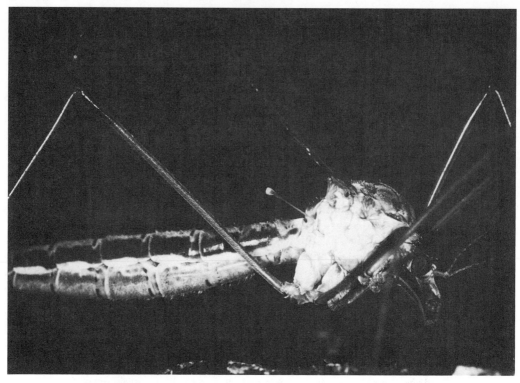

Figure 46. Lateral view of a crane fly (*Tipula dorsimacula*). Note the presence of the clubbed halteres.

In other insects, the forewings become the primary organs of flight with the hind wings being reduced, lost, or highly modified (Figs. 46, 47). When reduced but functional, hind wings usually are coupled to the forewings to act as a single unit for more efficient flight. The coupling may be by folds, hooks (Fig. 48), bristles, or enlarged lobes from either wing. Extreme modification of the metathoracic wing is seen in the fly in which it becomes a balancing structure of *haltere,* a vital role for an insect with a stout body and narrow wings.

In addition to being used in dispersal and locating food, wings also protect the body against physical damage, display a distinctive color pattern for mate attraction, and produce sounds. In some aquatic insects, wings also provide a covering to trap a reservoir of air. The wings are also colored to assist the insect in hiding or warning predators of their unpalatability. In some social insects, wings may be used in cooling nests. These latter functions will be discussed in future chapters.

Major Types of Legs

Ambulatorial. This is the least specialized and is often referred to as a *walking leg.* It consists of six segments (Fig. 49): the *coxa, trochanter, femur, tibia, tarsus,* and *pretarsus.* The femur and tibia are longer than the other segments and have a conspicuous "knee" between them that permits the insect

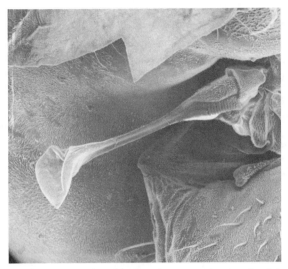

Figure 47. Scanning electron microscope photograph of the flesh fly *(Sarcophaga bullata)* haltere.

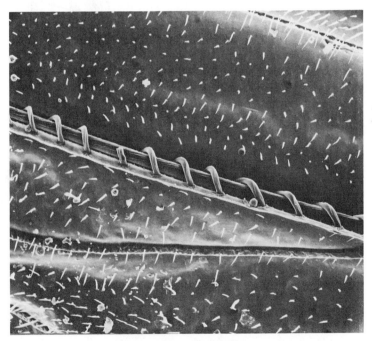

Figure 48. Scanning electron microscope photograph of the hooks from the hindwing, which connect to the forewings in a social wasp *(Polistes* sp.), producing a single functional unit for flight.

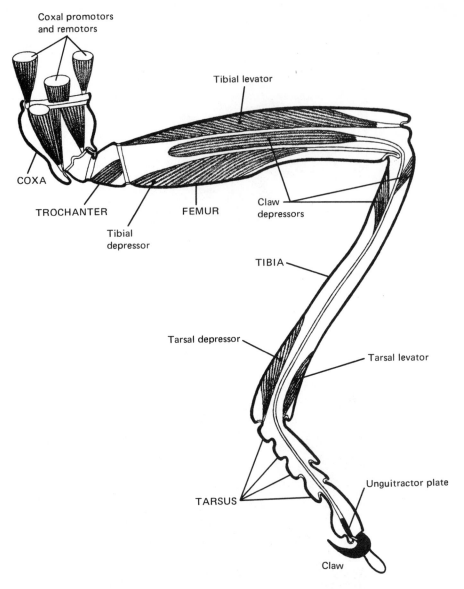

Figure 49. Segmentation and major musculature of a walking leg. (Redrawn with slight modifications from Snodgrass, 1935.)

to be slung low to the ground for stability. Although the tarsus appears to have segments, these segments are actually pseudosegments or *tarsomeres* since each lacks independent musculature. The pretarsus consists of only the claws or *ungues* and often either a single lobelike *arolium* or the two lobed *pulvilli*. Claws enable insects to move on rough surfaces. The aroleum and pulvilli or other types of adhesive pads and hair on the tarsi assist in movement across smooth surfaces.

The entire leg is moved by muscles originating on the tergum or sternum and inserted on the coxa. In addition, movement on individual segments is accomplished by muscles within the leg segments (Fig. 49). These muscles either extend or flex, thereby either pushing or pulling as the entire appendage is moved basally. In some insects, these muscles also suspend the body above the substrate while other insects permit their bodies to contact the surface when the insects are not active.

In walking or running, the front and hind legs on one side and the middle leg of the opposite side are raised and moved forward together. Once these legs have completed their movement, the other three legs are moved. As a result, the insect is always well-balanced and supported by a tripod of legs. Exceptions to the previous include insects that have highly modified front legs, such as the mantids during slow walking, and insects that have lost legs; in such examples, a tripod is maintained by moving only one leg at a time or else a modified behavior is evoked in which a staggering movement occurs during rapid locomotion.

Cursorial.   Most running animals have legs that are elongate and slim. Increased length permits greater distances to be covered with the same muscular effort, and the slimness reduces environmental friction. Tiger beetles and cockroaches are examples of insects that have this modification.

Saltatorial.   To saltate means to jump or vault. Legs modified for this function (Fig. 50A) commonly have greatly enlarged femora to accommodate the enlarged extensor muscles that straighten out the tibia. Because the legs are well anchored by large tarsal pads, claws, and often spines, a rapid contraction results in the entire body's being propelled (Fig. 51). Most legs of this type are located on the metathorax, as seen in the grasshopper, so that the direction of jumping is forward where the major sensory structures can perceive the upcoming environment.

Raptorial.   The front pair of legs is often modified to grasp and hold prey for feeding (Fig. 50B). The large muscles here are flexors and the tibia is pulled back against the femur when they contract. Spines may also be present on the femur and tibia, as in the mantid, to impale the prey and decrease the likelihood of escape by the victim.

In a *few* parasitic wasps the hind legs are raptorial. They hold the host near the stinging ovipositor and where the egg can be deposited. Modification of all legs to hold hair of the host may be seen in the sucking lice.

Natatorial.   Most of us are familiar with oars and swimmers' arm movements. The principles in their use are similar to the activities and modifications seen in the swimming legs of insects. Diving beetles, for example, have the middle and hind legs flattened, and the segments often are approximately the same size. When they are straightened and rapidly moved posteriorly, the maximal surface area is exposed to the water. Friction is further increased by the rows of strong setae on legs (Fig. 50D). The legs are

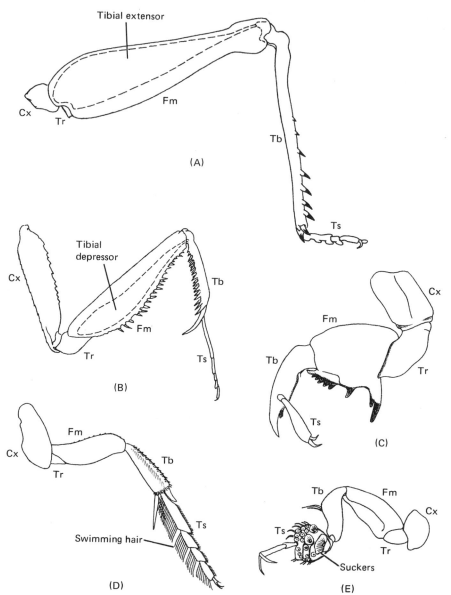

Figure 50. Modifications of the insect leg. (A) saltatorial (hind leg of grasshopper); (B) raptorial (front leg of mantid); (C) fossorial (front leg of cicada numph); (D) natatorial (middle and hind legs of water scavenger beetles); (E) clasping (front leg of male predaceous diving beetle).

returned more slowly and the legs are rotated so that the thin sides are exposed during the recovery stroke. Swimming hairs also fold back during this maneuver. The net effect is to propel the organism forward.

Fossorial. The forelegs of the mole cricket and cicada nymph (Fig. 50C) are shortened and heavily sclerotized. Large toothed projections from either the femur or tibia are used to "rake" through the soil to dislodge soil particles. The tarsi, as in raptorial legs, are reduced and usually fold back out of the way during excavation activity.

Clasping.   The forelegs of certain aquatic beetles (Fig. 50E) are modified for holding the female during copulation. Several tarsomeres are usually enlarged with suckers and large claws to produce these effective holdfast organs.

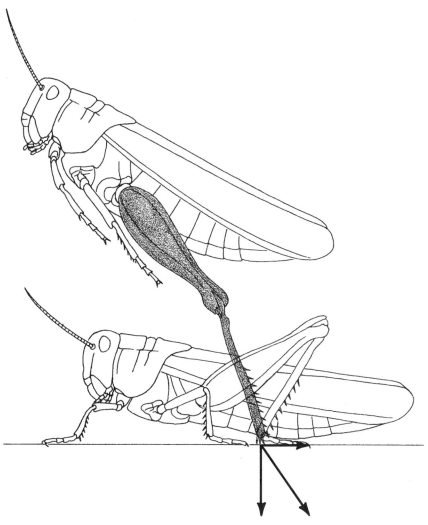

Figure 51.   Diagram of grasshopper jumping, showing thrust exerted by hind leg and its vertical and horizontal components. (Redrawn from R. F. Chapman, *The Insects, Structure and Function,* American Elsevier, 1969.)

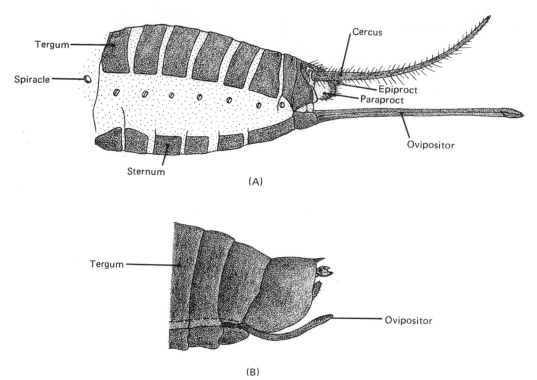

Figure 52. Basic structure of the abdomen with modifications for oviposition. (A) cricket; (B) cicada.

THE ABDOMEN

The posterior body region in insects is called the *abdomen.* The number of segments varies from 9 to 11 except in Collembola. Terminal segments often fuse or are reduced in size. The first segment may fuse with the thorax. The remaining segments, however, are very similar and consist of a dorsal *tergum* and a ventral *sternum* (Fig. 52). The anterior 8 segments usually have a pair of spiracles. Variation from the adult plan may be seen in some immatures in which sclerites may be lost or modified into many small plates and the number of spiracles vary greatly. The functions of this region are vital to the organism. It is in this region that the major viscera, heart, and reproductive organs are located. The reproductive openings and genitalia are found on the ninth abdominal segment in males and on the eighth and ninth abdominal segments in females.

External Genitalia

Male. In most terrestrial organisms, intromittent structures are present for penetrating the body of the female and depositing sperm. In insects, this *aedeagus* is often sclerotized in varying degrees and modified sufficiently to

50

prevent interspecies mating. The aedeagus may be paired (mayflies) or many-lobed (cockroaches), but it is usually a single phallus. In addition, a pair of *claspers* may be present to maintain the correct positioning of the female during copulation. In most insects the aedeagus and claspers are usually considered to be structures of the ninth abdominal segment, but there is some confusion as to their origin and whether they represent modified appendages or secondary development from sternal lobes.

Female.   Correlated with copulation (to be discussed under reproduction) is the need to deposit eggs. In the primitive forms, eggs were undoubtedly just "dropped," but most present-day insects have an ovipositor of some sort for placing and positioning the eggs in an appropriate microhabitat. The most primitive type, seen in grasshoppers (Fig. 140) and cicadas (Fig. 52), consists of a sclerotized tube formed from the paired appendages of both the eighth and ninth abdominal segments. The shape, degree of hardness, and length of this structure usually determine where the eggs are to be placed, e.g., in the ground, under bark, etc. In a few insects such as bees and wasps, this ovipositor has been further modified into a sting (Fig. 53) and the primary role of egg deposition has been lost.

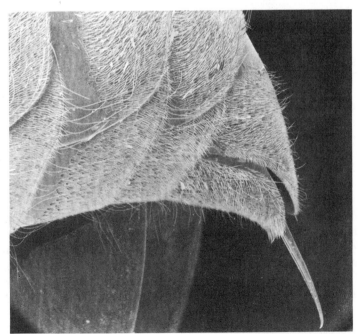

Figure 53. Abdomen of a vespid wasp (*Polistes* sp.) with the terminal sclerites parted and sting thrust into stinging position as seen through the scanning electron microscope.

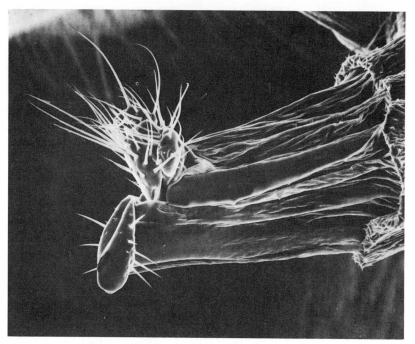

Figure 54. Scanning electron microscope photograph of the terminal end of the tubular ovipositor everted from the face fly (*Musca autumnalis*). Note the numerous sensory setae, particularly at the terminus.

The more advanced insects, such as flies, have lost their appendicular structure, and another type of ovipositor consisting of a tube formed from the terminal abdominal segments has evolved (Fig. 54). Under normal circumstances it remains telescoped within the anterior segments, but when blood pressure is increased, it is extruded and capable of depositing eggs. Various sense receptors are located at the terminus (Fig. 54) and each segment has retained muscles for movement to permit very exact positioning of the ovipositor.

Cerci

Embryologically, each segment bears a pair of appendages but these are usually lost in the abdomen prior to hatching from the egg. The eleventh abdominal segment, however, often retains its appendages, the *cerci*. In the most primitive species (Fig. 229) they are long and multisegmented and have numerous tactile sensory setae. These are especially advantageous in soil dwellers because they permit sensory pickup when the insects move backward. In the more advanced insects, these structures tend to become reduced or lost.

Larvapods

Retention and development of abdominal appendages in immatures may occur. For convenience, these appendages are collectively termed *larvapods*.

These adaptations are lost when metamorphosis occurs, but their importance in the life cycle should never be minimized.

Some larvapods, such as mayfly nymphs, are modified into tracheal gills (Fig. 55C) and are found on most abdominal segments. It is an interesting experience to sit and watch as these gills are vibrated thereby increasing the water flow across these respiratory organs.

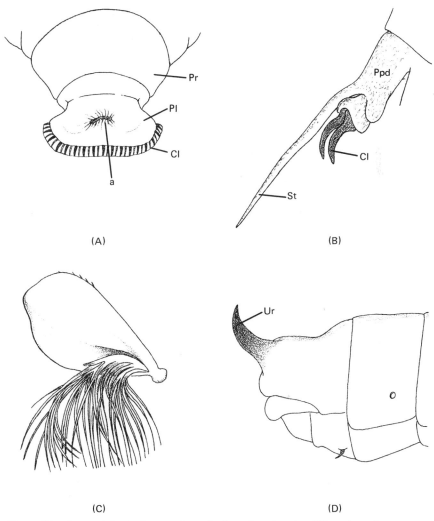

Figure 55. Some abdominal appendages in immature insects. (A) proleg of lepidopteran caterpillar with numerous hooks or crochets for grasping rough surfaces and a suckerlike area (a) for smooth surfaces; (B) pygopod (appendage of the 10th abdominal segment) of dobsonfly larva for grasping rocks and maintaining position in swift water; (C) gill of mayfly nymph; (D) urogomphus (appendage of 9th abdominal segment in larvae) of ground beetle; Cl, claw; Pl, planta; Ppd, pygopod; Pr, proleg; St, stylus; Ur, urogomphus.

Other aquatic organisms, such as caddisfly larvae, have hooked holdfasts (Fig. 55B) at the posterior end of their bodies which enable them to retain their position on rocks or to hang on to a portable "house" or case (Fig. 280). These appendages are most often noted in species found in moving water.

There are few individuals who have not observed creeping caterpillars and noticed the lobelike *prolegs* that assist the caterpillars in locomotion. In lepidopteran species, little hooks or *crochets* (Fig. 55A) are present to permit tenacious gripping on rough substrates. When smooth surfaces are encountered, the region with hooks is rotated dorsally to expose a suckerlike area for increased efficiency in locomotion.

Some larvae possess appendages on the ninth segment, the *urogomphi* (Fig. 55-D). The development of these varies from cercilike to fixed horny outgrowths.

Questions

1. What are the major functions of an insect head? How do these compare with those of other animal heads?
2. What insects would be expected to have the most sclerotized heads? Why?
3. What appendages are associated with the head?
4. What types of mouthparts are modified for fluid uptake? Are these found in the adult, larval, or nymphal stages?
5. Which of the mouthpart types are associated with the greatest range of diets?
6. Is there a correlation between diet and the number of ommatidia present in the compound eye? Elaborate.
7. How many segments make up the thoracic region? Which of these is or are the most highly modified? Does any change in the thorax occur in progressing from an immature to an adult?
8. Are wings found in all insects and all stages? How do wings develop?
9. Why cannot wings be appendages?
10. What is the difference between Neopterous wings and Paleopterous wings?
11. What advantages are gained by the presence of wings? What are some disadvantages?
12. Why is it only the forewings and not the hind wings that become protective structures?
13. What are some of the modifications of insect legs? How does each modification better equip the organism to survive?
14. How many segments are in the abdomen? Are they the same structure as the head and thoracic segments?
15. Are any appendages associated with the abdomen?
16. What is an ovipositor? What selective advantage does it give the possessor?

3

The Insect Internally

DIGESTIVE SYSTEM

Single-celled and some small multicellular animals obtain nutrients either by diffusion or by engulfing food particles. The exchange rate is *proportional to the amount of surface area.* Once inside the cell, the molecules provide building blocks necessary for growth (assimilation) or are directed into an energy cycle (respiration). In contrast to food uptake, the rate of metabolism is *proportional to cellular volume.* With continued growth, size eventually becomes limiting, for the volume, which determines metabolism, increases at a more rapid rate than the surface which supplies nutrients and oxygen. Various internal chambers and tubes, i.e., respiratory and digestive systems, have been incorporated in most animals to provide the necessary increase in surface area(Fig. 56).

Insects possess a complete tube or *alimentary canal* that takes in food through an anterior *mouth,* breaks down this food by enzymatic hydrolysis, absorbs nutrients into the body, and evacuates the remaining matter to the exterior through a posterior *anus.* Enzymes secreted are specific to the diet of the individual. Various glands to increase enzyme production have also evolved but not to the extent seen in vertebrates.

The insect alimentary canal (Fig. 57) develops from two invaginations, one anterior and one posterior, of the embryonic exoskeleton (ectoderm)

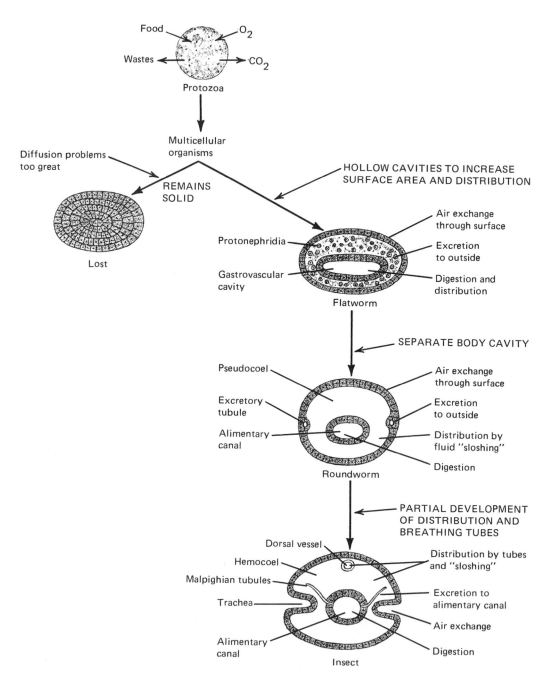

Figure 56. Diagram of means of increasing surface area to carry on the biological needs as organisms get larger and decrease the surface to volume ratio and hence movement of vital substances.

Figure 57. Embryonic development and specialization of the alimentary canal. (A) stomodeal inpouching occurs at anterior and proctodeal inpouching at the posterior; (B) primordial mesenteron forms at ends of inpouchings; (C) mesenteral elements unite to form blind sac; (D) regions between stomodeum, mesenteron, and proctodeum split to form continuous tube or alimentary canal. Specialization occurs to produce specific structures: AInt, anterior intestine; An, anus; Cr, crop; GCa, gastric caecum; Mal, malpighian tubule; Ment, mesenteron; Mth, mouth; Oe, esophagus; Phy, pharynx; Proc, proctodeum; Pvent, proventriculus; Py, pylorus; Rect, rectum; Stom, stomodeum; Vent, ventriculus. [Part (D) redrawn with slight modifications from Snodgrass, 1935]

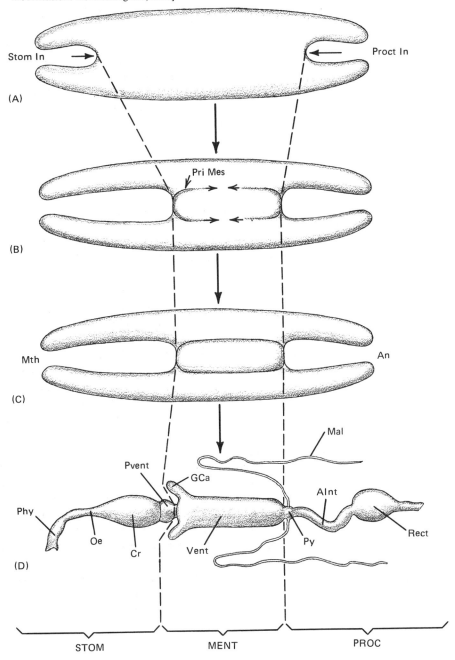

which grow toward and unite with a hollow tube formed from endoderm. The boundaries between these three sections break down to form a single, continuous tube. The anterior section or invagination is called the *stomodeum,* the posterior section or invagination is the *proctodeum,* and the central endodermal section is the *mesenteron.* Because of their ectodermal origin, both the stomodeum and proctodeum have a lining of cuticle, the *intima.* This cuticle is not found in the endodermal mesenteron.

Variation within each of the three sections of the alimentary canal depends upon diet (Fig. 58). Also, the digestive tract may change in insects that undergo complete metamorphosis to the adult stage with the result that an immature that is equipped to handle solid food may emerge as an adult feeding on nectar or blood (Fig. 59). This metamorphosis certainly must be classified as one of the marvels of nature.

Solid Food

Solid food requires grinding prior to ingestion. Grinding is normally carried out in a chamber, as it is in most higher animals. This chamber, or *preoral cavity* (Fig. 30), is formed by the anterior labrum, the lateral mandibles and maxillae, and the posterior labium. Food is chewed into small pieces, mixed with salivary secretions from *labial glands* for lubrication and very limited carbohydrate digestion, and then pushed into the mouth and swallowed. The muscular *pharynx* assists in initial food intake. Food is now moved down the tubular *esophagus* by waves of muscular contractions termed *peristalsis.* In primitive insects, such as springtails, the food bolus now enters the mesenteron; in more specialized insects, such as the grasshopper (Fig. 58), the posterior part of the stomodeum has been modified into an expansible *crop* for temporary food storage and a muscular *proventriculus* that acts primarily as a valve. Because of the presence of the intima, little to no food has been digested or absorbed except in a few forms, such as the cockroach, in which some enzymes may be regurgitated forward into the crop to effect varying degrees of digestion.

When the proventriculus relaxes, food is moved posteriorly into the mesenteron. The tube here is called the *ventriculus.* Lateral diverticula or *gastric caecae* may expand outward from the ventriculus to increase the surface area. It is in this region (mesenteron) that the major enzymes are secreted. Enzymes are liberated from the midgut epithelial cells by releasing vacuoles or by a complete breakdown of cells containing them. Once liberated, enzymes are activated and proteins are broken down to amino acids by a group of enzymes termed *proteinases.* Carbohydrates are acted upon by various enzymes collectively termed *carbohydrases* and broken down to simple sugars or monosaccharides. Fats and oils are broken down by *lipases* into fatty acids and glycerol.

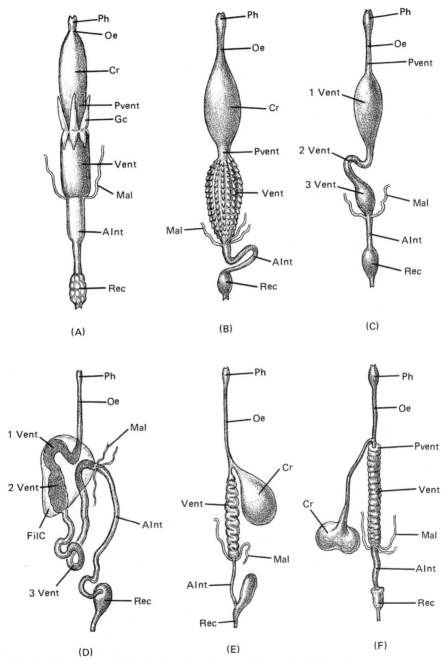

Figure 58. Some variations in the alimentary canal in insects. (A) grasshopper; (B) ground beetle; (C) water strider; (D) cicada; (E) moth; (F) housefly; AInt, anterior intestine; FilC, filter chamber; Gc, gastric caecum; Mal, malpighian tubules; Oe, esophagus; Ph, pharynx; Pvent, proventriculus; Rec, rectum; Vent, ventriculus.

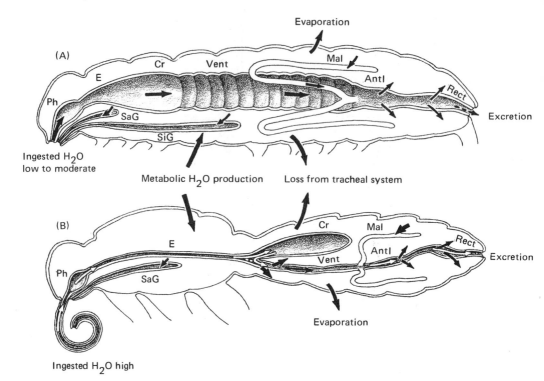

Figure 59. Movement of water into, around, and out of the body. (A) caterpillar ingesting solid food; (B) adult butterfly of the same species ingesting liquid food. Open arrows indicate high water content, black arrows moderate water content, and broken arrows low water content.

Correlated with the absence of the intima in this region is the absorption of food. Both the ventriculus and gastric caecae absorb the above products of digestion. The uptake may be either passive (diffusion) or require energy (active transport). Food is retained in the midgut until the digestive and absorptive periods are complete whereupon the pyloric valve of the hindgut relaxes and the remaining material is moved posteriorly into the anterior intestine.

The intima in the proctodeum is sufficiently permeable to permit the absorption and reabsorption of water and salts (Fig. 59). As the fecal material is moved posteriorly, it becomes progressively dehydrated until it is formed into a pellet in the rectum and subsequently evacuated.

The cellular lining of the alimentary canal is protected from abrasion by the intima of the stomodeum and proctodeum and the delicate *peritrophic membrane* in the mesenteron in some insects. The peritrophic membrane is formed either by cellular secretions or by the loosening of cellular borders of the mesenteron, but the peritrophic membrane remains permeable to enzymes and digestive products. The membrane is moved down the canal and often evacuated with the feces.

60

Liquid Food

Modifications of the digestive system for liquid diets are many and usually correlated with a corresponding change in mouthparts. The most common structural modifications are the narrowing of the tube and the presence of a diverticulum or *crop* off the esophagus for storage of liquids (Fig. 59B). Excess water, in insects that ingest liquids, often becomes a problem, especially in immatures in which large amounts of food must be taken in to facilitate growth; this excess water must be eliminated for osmoregulatory reasons. Excess water also makes the enzyme–substrate encounter more difficult because of the increased distance between food particles.

Another modification, as seen in Fig. 58C, results in sections of the ventriculus becoming morphologically distinct and named ventriculus 1, 2, 3, and rarely 4. Little is known about physiological functions of each, however.

Some insects must ingest excessive liquids to obtain minimal amounts of nutrients. This is particularly true of sap feeders. One such group, the Homoptera (Fig. 58D), has solved this by concentrating the food in a *filter chamber*. An anterior section and a posterior section of the alimentary canal touch one another in an enclosed sac in which much water is short-circuited between these sections and evacuated to the exterior. The net effect is food concentration for efficient digestion in the ventriculus without disrupting the osmoregulation system of the body.

Of special importance are the enzymatic changes that occur in the evolution to liquid feeding. The labial glands and regurgitation of ventricular fluids become exceedingly important and produce anticoagulants in most blood feeders, toxins in many carnivores, histolyzing enzymes in many herbivores and carnivores, etc. The change in diet also results in varied enzymatic changes in the digestive materials secreted by the mesenteron and by the microorganisms that may exist symbiotically in the gut or gut lining.

Symbionts

Many insects have cultures of microorganisms within their body which provide their host with vital nutrients and the capacity to utilize many nutrient deficient foods. Many are found intracellularly throughout the body, but we will restrict our discussion to those found in the gut.

Although many insects ingest cellulose, few have the enzyme *cellulase* to digest this complex polysaccharide. Digestion of this carbohydrate is normally accomplished by bacteria and protozoa located in the alimentary canal of the host. Termites, for example, feed on fecal material and oral feedings from older members of the colony after hatching from the egg and are thereby inoculated with the microorganisms necessary to digest the very food they feed upon.

Some insects may have specialized chambers to "house" these organisms. If these chambers are part of the hindgut as in termites, the lining of this region along with the symbionts are shed at each molt. Reinoculation must then occur for survival.

CIRCULATORY SYSTEM

With multicellular organisms and their increased size, distribution of metabolic materials becomes a problem to be first solved by a body cavity filled with "sloshing" fluids (Fig. 56) and later by incorporating distribution tubes into this system. When tubes or vessels are absent or when they are present but the fluid is not completely restricted within them, as in insects, the system is referred to as *open* (Fig. 60). When the fluids are confined within vessels, the system is said to be *closed*.

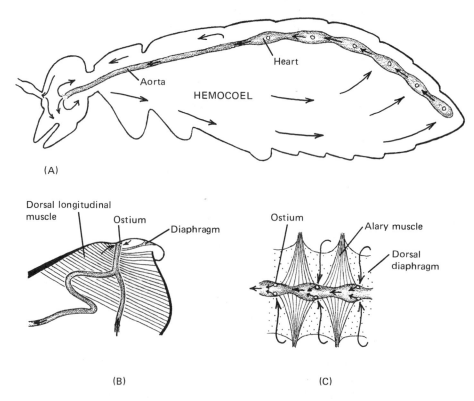

Figure 60. Circulatory system structure. (A) lateral view of hypothetical insect indicating directional blood flow; (B) accessory pulsating organ; (C) dorsal view of heart section. [Part (C) redrawn with slight modifications from Snodgrass, 1935]

Insect blood, or *hemolymph,* comprising from 5 to 40 percent of the body, is enclosed within the body cavity or *hemocoel* (Fig. 60A). A longitudinal *dorsal vessel* is present in most insects and is divided into a series of pumping chambers (*heart*) in the abdomen and the *aorta* which extends forward to the brain region (Fig. 60C). Hemolymph enters each chamber through heart openings, the *ostia,* and is pumped forward by peristaltic contractions of the heart to the brain where it flows out into the hemocoel. Blood then generally flows posteriorly along the alimentary canal where the absorbed nutrients are picked up from the mesenteron for distribution to the body. The hemolymph now flows dorsally through openings in the *dorsal diaphragm,* which suspends the dorsal vessel, and reenters the heart (Fig. 60A).

The previous description applies only to small resting insects. Larger resting insects often require breathing movements, consisting of abdominal segment contractions, to assist in fluid distribution. Active insects "slosh" the hemolymph about during locomotion movements, especially flight, and this aids the mixing action and hemolymph movement. Efficiency by exploiting contractions of locomotory muscles is sometimes enhanced by specialized *accessory pulsating organs* (Fig. 60B).

Blood cells, or *hematocytes,* are present in the hemolymph but do not carry oxygen. Their function is usually similar to the white cells of humans and mainly are phagocytic.

In summary, the circulatory system functions in transportation of nutrients, hormones, etc., about the body but has little to do in oxygen transfer. Of less conspicuous but of perhaps equal importance is its role in keeping cells moist and maintaining osmotic pressures, regulating heat within the body, buffering or detoxifying reactive molecules, healing wounds, protecting against foreign invaders, and maintaining blood pressure that assists in molting and in locomotion, particularly for legless forms.

RESPIRATORY SYSTEM

The word respiration has acquired several different meanings; to some it indicates the exchange of air or ventilation while to others it refers to the breakdown of food to obtain energy. Modern physiologists now use the word to indicate a series of biochemical reactions occurring in cells that liberate energy. In the presence of oxygen, the reactions may be summarized as:

$$(1) \qquad 6O_2 + 6H_2O + C_6H_{12}O_6 \longrightarrow 6CO_2 + 12H_2O + \text{energy}$$

The above equation is an extremely simplified one in which the initial compound is a simple sugar or monosaccharide. Monosaccharides come from the breakdown of cellular glycogen and the blood sugar trehalose. Other substances, such as amino acids and fatty acids, may also be shunted into this energy cycle in various places. A more detailed summary of the reactions would be:

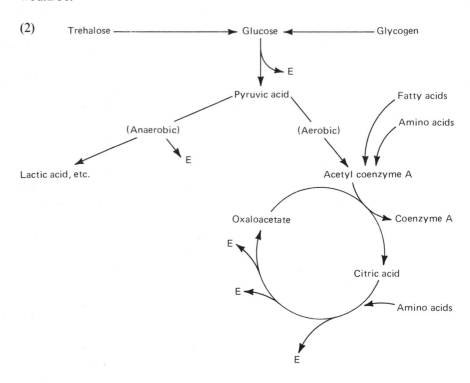

(2)

The liberated energy comes at many places in the above summary and is most commonly used to convert adenosine diphosphate (ADP) to adenosine triphosphate (ATP) which in turn serves as a ready source of high energy.

(3)   Energy + ADP + inorganic phosphate ———————► ATP

Approximately 55 percent, or 380,000 calories, of the total energy of the sugar molecule is released and stored in the resulting 38 molecules of ATP when going through the oxidative cycle (*aerobic respiration*). Carbohydrates and protein respiration result in similar amounts of energy (4 calories/g), but fats are more productive (9 calories/g). If oxygen is not available, the food passes into another metabolic pathway (*anaerobic respiration*) yielding only a small fraction of the potential energy. Since only 2 molecules of ATP result from anaerobic means, nearly 20 times as much glucose must be metabolized to obtain as much energy as through the aerobic form.

Insects are cold-blooded and the respiratory rate is usually proportional to the temperature of the external environment. Species in the tropics have lit-

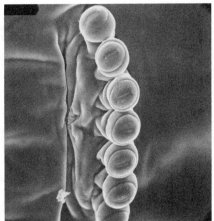

Figure 61. Prothoracic spiracles of adult flesh fly (*Sacophaga bullata*) showing filtering setae, which prevent dust and foreign particles from entering the tracheal system.

Figure 62. The prothoracic spiracle of a house fly (*Musca domestica*) larva. This spiracle, as seen through the scanning electron microscope, has an opening at the tip of each of the finger-like processes that unite internally into a common atrium.

tle difficulty in maintaining an active metabolism. For those surviving in cold regions of the world, however, nearly total inactivity occurs for significant periods of their life. This inactivity is advantageous, especially during the winter for herbivorous species, because their food source is not available and energy would be wasted by fruitless searching and activity.

The membranes of any respiratory organ must remain moist in order to permit oxygen to dissolve and pass through. This is no problem for small aquatic species because of their habitat and large surface/volume ratio but terrestial insects, because of selection for exoskeletons that control water loss, have this route limited. Portions of the exoskeleton, however, have been invaginated into the body to form an extensive separate system, up to 50 percent of the total body volume, of respiratory tubes, the *tracheae,* which carry oxygen directly to tissues. Blood or hemolymph normally plays no role in oxygen transfer as it does in other animals.

Most insects have openings or *spiracles* into the tracheae. The system is referred to as *open.* Spiracles vary from simple holes to highly modified structures having filters and valves for regulating the openings (Figs. 61, 62). Most species have two pairs of thoracic and eight pairs of abdominal spiracles. Each (Fig. 63) sends a branch dorsally to muscles and the dorsal vessel, medially to the alimentary canal and gonads and one ventrally to muscles and the nerve cord. Tracheae are prevented from collapsing by spiral braces (*taenidia*) in

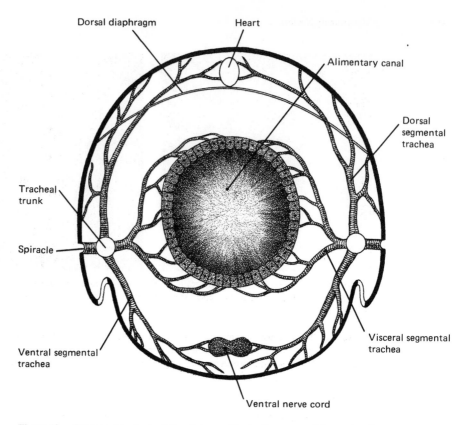

Figure 63. Cross section through the abdomen illustrating some of the tracheation. (Redrawn with slight modifications from Snodgrass, 1935)

their wall and each progressively branches into smaller tracheae until minute *tracheoles* (nontapering tubes) are reached. The radiating tracheole network resembles the capillary network in humans, and it is through these structures with their extensive surface area that most gaseous transfer occurs.

What about the body segments that do not have spiracles, such as the head? What happens when a spiracle becomes covered or blocked? All segments are interconnected by longitudinal *tracheal trunks* as seen in Figure 64. Oxygen enters the segments that have spiracles and moves throughout the body to segments lacking spiracles by diffusion plus muscular movements of the abdomen.

Air sacs of various sizes may be present and are recognized by their reduction or lack of braces. Their functions are (1) to serve as reservoirs of oxygen, (2) to serve as bellows in distributing air and cooling the body, particularly during flight, (3) to decrease weight in fast-flying species, and (4) to increase body pressure during certain periods such as molting.

Lack of oxygen, excess of carbon dioxide, or both are the normal initiators of spiracular and ventilatory activiy. Having the spiracles closed except

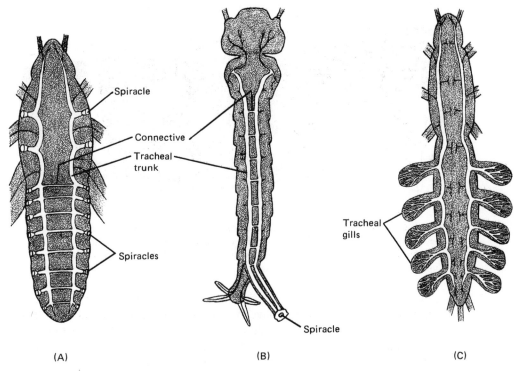

Figure 64. Some variation on the tracheal system of insects. (A) typical type of open system; (B) open system with reduction in number of spiracles; (C) closed system with cutaneous and tracheal gill gas exchange.

for brief periods of breathing results in the capability to control water loss more efficiently. The ability to close the spiracles and having a hard exoskeleton and air reservoirs internally are beneficial during certain periods of environmental stress. For example, attempts to control grain beetles by fumigation are successful only after long periods of exposure which result in oxygen depletion, the system opening, and the active agent being inhaled.

A few tracheal systems are *closed* and gaseous movement into the tracheae for distribution must first pass through the exoskeleton proper. Some aquatic insects have increased surface area in the form of tracheal gills, evaginations of the exoskeleton with extensive tracheole networks (Fig. 64C).

MUSCULAR SYSTEM

Although most cells are capable of limited contraction, certain cells have great ability to contract. The degree of contractibility and the rate is determined by intracellular bands or *striae* that carry out the actual process; those

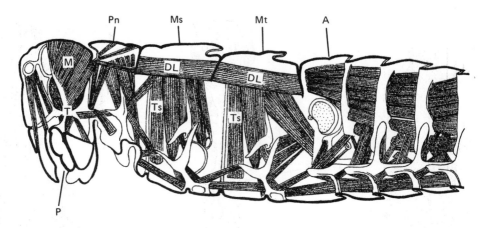

Figure 65. Musculature of a grasshopper (*Melanoplus differentialis*). A, abdominal segment; DL, dorsal longitudinal muscle; M, mandibular muscle; Ms, mesonotum; Mt, metanotum; P, preoral cavity; Pn, pronotum; T, tentorium; Ts, tergo-sternal muscle.

muscle cells with relatively few striae are characterized by a slow rhythmic contraction while vigorous ones have extensive networks. Unlike the muscle cells in mammals, all insect muscle cells have striae, but the number varies greatly.

Muscle cells are aggregated into muscle fibers and in turn into functional units called *muscles*. The number of individual muscles in an insect varies markedly. The adult grasshopper (Fig. 65) has approximately 900 muscles and butterfly caterpillars have approximately 4,000 (humans have 800).

The power of a muscle is proportional to the area of its cross section. Since the mass that must be moved is proportional to its volume and because of somewhat better leverage systems than humans, small insects are able to carry out what appear to be miraculous performances. Fleas, aided by special elasticity in the pleuron, are capable of leaping distances that would approximate a human's leaping more than 1,000 ft (305 m). Ants are capable of moving weights that would approximate several tons when the size and weight are proportionately increased to compare with that of a human. These examples, however, are only valid for comparisons and cannot be construed to indicate potentials since size is such a limiting factor.

Muscles in insects are often artificially segregated into four types: *visceral, segmental, appendicular,* and *flight*. Visceral muscles surround ducts and tubes and produce directional waves or peristalsis to move products from one region to another. Segmental muscles cause telescoping of segments necessary in molting, inhalation and exhalation, increasing body pressure, and locomotion in legless individuals. Appendages are moved as a unit by muscles originating on either the tergum or sternum and inserted on the coxae. In addition, as previously discussed, each segment is operated by muscles originating in the previous segments (Fig. 49).

It is in the flight muscles that one sees the greatest specializations. One type is termed *synchronous* and produces one contraction per nerve impulse. Because of this limit, the rate of wing movement is restricted to a maximum of approximately from 30 to 40 beats/s with most species in the 5 to 15 beats/s range. Another type is *asynchronous* and is found in highly specialized insects, such as flies, and has an innate contraction rhythm much faster than the nerve impulse rate. Some gnats and mosquitoes move their wings from 500 to 1,000 beats/s, which is fast enough to produce air vibrations that humans can hear.

Such accelerated contractions obviously consume great amounts of energy, approximately 1,000 calories per gram of muscle per hour. This is from 10 to 20 times greater than vigorous exercise in humans and is only possible because insects are so small. The energy comes originally from glycogen or fat, the latter being most efficient (8 times) and common, especially in insects that fly great distances. These food sources provide the energy necessary to convert ADP to ATP in the numerous mitochondria, to 40 percent of the volume, found in insect muscle cells.

Since insects are cold-blooded their temperature varies directly with the environment. This variation is beneficial to the small insect, for great amounts of energy loss (more than 90 percent) would be required to metabolically burn sufficient food to maintain a uniform body temperature with such a high sur-

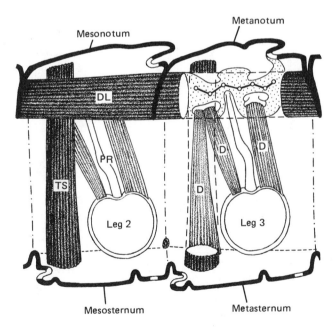

Figure 66. Meso- and metathorax of a grasshopper with all but the flight muscles removed. D, direct flight muscles; DL, dorsal longitudinal muscles; PR, pleural ridge; TS, tergo-sternal muscles.

70

Downstroke

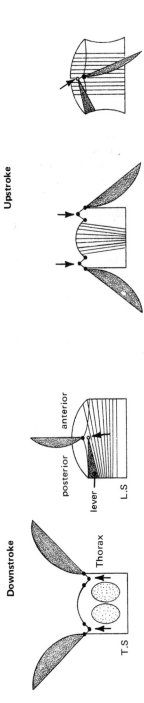

Figure 67. The click mechanism of a fly's wing. Wing movement is driven by muscles that distort the thorax. (Redrawn from Wells, *Lower Animals*, 1968)

Upstroke

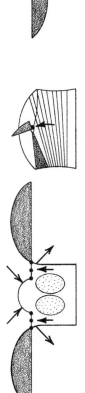

(A) Longitudinal muscles contract. Scutellar lever (arrowed) is forced upwards.

(D) Dorsoventral muscles contract. Scutellar lever (arrowed) is forced downwards.

(B) Thorax sprung; lever passes midpoint

(E) Sides and top of thorax sprung

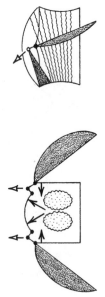

(C) Sudden relaxation of tension on longitudinal muscles as thorax springs back into shape.

(F) Scutellar lever passes midpoint. Thorax springs back suddenly removing tension from muscles.

face/volume ratio. However, the temperature of flight muscles must normally approach from 30° C to 35° C for flight to occur. How is this possible? In some insects this temperature is not maintained and flight becomes greatly restricted. Some insects expose themselves to maximal sunlight whereby the body heats up sufficiently for flights. However, in some of the larger species whose surface/volume ratio is too low or when the insects are nocturnal, sufficient warming cannot occur by solar radiation. These require a warm-up period of wing vibrations to generate sufficient metabolic heat to raise the muscle temperatures into the flight range. In cool weather this may require from 10 to 15 minutes. To prevent excessive temperature buildup during flight, heat is transferred to the abdomen by the blood and is lost through radiation to the air, either through the air sacs during breathing or from the external exoskeleton.

The mechanics of flight differ from the other animals capable of flight (bats and birds) because the wings are not modified appendages. The movement of the wings usually results *indirectly* from the distortion of the tergum. Because the pleural wing process serves as a fulcrum, the contraction of the *tergo–sternal muscles* (Fig. 66) lowers the tergum, which in turn raises the wing tips. Lowering the wing tips results when the tergum is vaulted upward as a result of the contraction of the *dorsal longitudinal muscles.* A complex click mechanism (Fig. 67) greatly amplifies the power of each stroke because of the sudden release of tension built up in the thoracic plates as the wing clicks into a new position of stability. The shape and rigidity of the thorax plays an important role in determining how this system functions.

The simple movement of wings up and down, however, does not produce the flight as seen in insects. Directional movement is accomplished by contraction of *direct flight muscles* (Fig. 66) which raise and lower the front and back edges of the wings. The combined action of the indirect and direct flight muscles is, therefore, to push air down and backward and the insect is drawn forward into the resulting region of low pressure. Further changes in wing pitch determine directional changes, elevation, and backward movements.

Although insects are able to move their wings rapidly, they are, nevertheless, slow fliers. The maximal speed known, based on experimentation rather then speculation, is approximately 36 mph (58 km/h) by a dragonfly. Average speed, however, is probably in the 5 to 10 mph (8 to 16 km/h) range.

EXCRETORY SYSTEM

As the direct result of metabolism within the cell, certain molecules, such as protein wastes, must be eliminated. Single-celled organisms excrete wastes directly into the external environment with little or no change, but multicellular organisms usually must detoxify wastes by combining them with other substances unless copious amounts of water are present to dilute them.

Specialized organs and ducts are present in most animals to remove these products from the blood and eliminate them from the body.

The major nitrogenous waste product in insects is *uric acid,* a relatively nontoxic and highly insoluble molecule. Since large amounts of water must be present to maintain uric acid in solution, the excretory organs or *malpighian tubules,* which extract it from the blood, open into the *anterior intestine* where the water and salts can be removed for recycling and the resulting uric acid crystals are carried to the external environment with feces. Some uric acid, however, is incorporated into tissues and the exoskeleton. In some cockroaches, up to 10 percent of the total dry body weight consists of uric acid.

As in other animals, excesses of any type are eliminated in order to prevent osmotic imbalances from occurring. Insects that ingest much liquid food (Fig. 59), and some freshwater inhabiting insects (Fig. 105), must eliminate water in order to maintain proper osmotic balance. During these periods of excesses, water is removed from the blood by the malpighian tubules and little or no absorption occurs in the proctodeum. (See Table 1.)

TABLE 1. Representative Numbers of Malpighian Tubules as Related to Metamorphosis

No Metamorphosis		Incomplete Metamorphosis		Complete Metamorphosis	
Order	*Number*	*Order*	*Number*	*Order*	*Number*
Collembola	0	Ephemeroptera	40–100	Neuroptera	6–8
Thysanura	6	Odonata	50–60	Hymenoptera	12–150
		Orthoptera	2–12	Lepidoptera	6
		Plecoptera	50–60	Mecoptera	6
		Hemiptera	0–4	Trichoptera	6
				Diptera	4

In addition to the malpighian tubules, various phagocytic cells and the fat body play an important role in the elimination or detoxification of waste products from the circulating blood.

NERVOUS SYSTEM

All cells are *irritable,* i.e., are capable of response to stimuli. In multicellular organisms certain cells have become specialized to carry these responses and to coordinate the incoming information into behavioral action. The potential number of these *neurons* is usually restricted by the size and specialization of the organism. Insects are relatively small and have a restricted

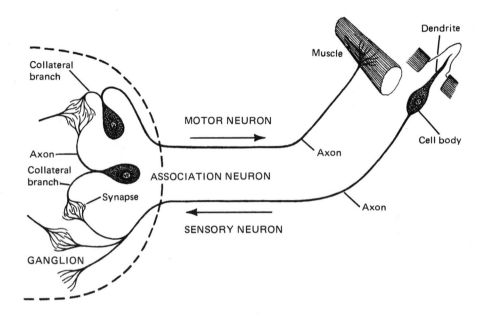

Figure 68. Diagram of the reflex mechanism of the insect nervous system. Note that it is opposite to the chordate type in vertebrates, where the sensory neurons enter the dorsal part of the ganglion and the motor leaves ventrally. (Redrawn with slight modifications from Imms, Richards, & Davies, 1957)

number of neurons. Nevertheless, these few neurons are used very efficiently through a series of "built-in" or innate behavioral patterns.

Neurons are of two basic types: In one, the *sensory* neuron, there are one to many dendrites. The second type, the *motor* and *association neurons,* lacks dendrites. Impulses in motor and association neurons travel only along the axon, originating along a *collateral branch* and normally do not pass through the cell body (Fig. 68). In sensory neurons the impulse is passed along the dendrite to the cell body before reaching the axon.

The nerve impulse starts at some sensory structure and represents an ionic change as depolarization of the membrane passes progressively along the cell. Because of the length of neurons, impulses are carried more quickly and efficiently than if the message had to be passed through normal-sized cells. Between two nerve cells (synapse), specific chemicals or neurotransmitters are released by the axon of one that initiates an impulse in the second.

Nervous tissue arises early during embryological development and becomes segmented as the individual metameres are formed (Fig. 69). These neural tissues form paired ganglia in each segment and are the bases of the

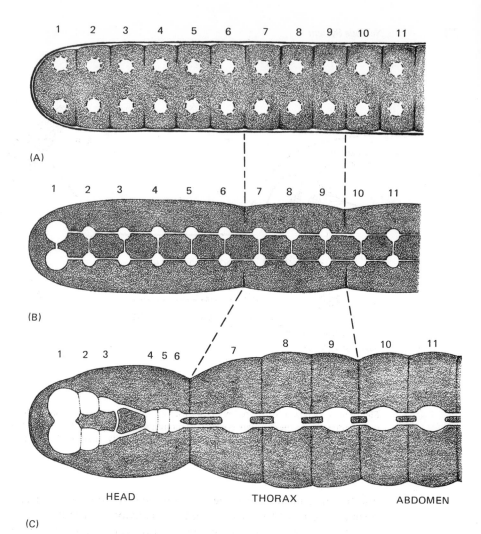

Figure 69. Diagrammatic representation of the development of the central nervous system. Nerves to the periphery have not been included. (A) neuroblast formation during segmentation; (B) interconnection of ganglia; (C) differentiation of ganglia as tagmosis occurs.

central nervous system. Ganglia become interconnected as the neuron fibers grow from one ganglion to another giving the central nervous system a ladderlike appearance. The position of the most ganglia is opposite to that of humans, being located below the digestive system; each pair of ganglia coordinates the activities of structures of the segment in which it formed. Some of the cells in ganglia secrete hormones and are called *neurosecretory* (see Chap. 4).

The ladderlike appearance is only relative, however. The anterior three pairs of ganglia fuse to form the brain or *supraesophageal ganglion* and the fourth to sixth pairs unite into the *subesophageal ganglion*. The remainder of the central nervous system is termed the *ventral nerve cord* and these ganglia also tend to fuse, especially each segmental pair and the last three to four pairs in the abdomen. The primitive origin of each pair of ganglia can be determined by tracing nerves to the segments they innervate.

The supraesophageal ganglion, as the name indicates, is located dorsal to the esophagus (Fig. 70). The first pair of lobes, the *protocerebrum*, receive nerves from the compound eyes and ocelli. It is the major association region in the central nervous system and its direct connection to the photoreceptors indicates the great effect light stimuli have upon most insects. Also, experiments on social insects have shown that the comparative size of this area seems to be correlated to the ability to learn. The *deutocerebrum* or second pair of ganglionic lobes receives impulses from the antennae, coordinates this sensory input, and controls the movement of the antennae. The *tritocerebrum,* unlike the other sections of the brain, remains separated into two lobes and receives nerves from the frontal ganglion, labrum, and subesophageal ganglion. All lobes of the brain are interconnected through nerve fiber tracts.

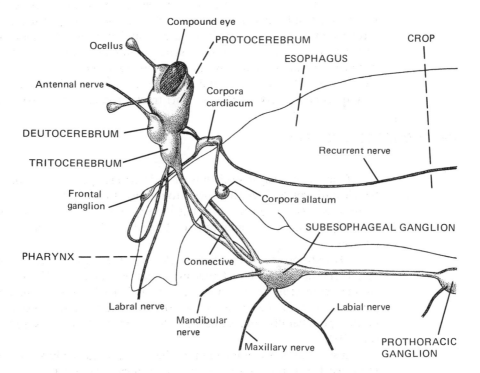

Figure 70.   Insect brain and associated structures. (Redrawn with slight modifications from Snodgrass, 1935)

The subesophageal ganglion is located below the esophagus and coordinates the sensory and motor activities of the appendages of the fourth, fifth, and sixth segments (the mandible, maxillae, and labium, respectively). Nerves also supply the hypopharynx and salivary glands.

The *stomodeal sympathetic system* is connected to the central nervous system through the tritocerebrum. It commonly consists of a single *frontal ganglion* plus one or two recurrent nerves along the dorsal surface of the stomodeum. In addition to controlling the peristaltic movements of the anterior part of the alimentary canal, this system also sends nerves to the dorsal vessel and two pairs of endocrine glands, the *corpora allata* and *corpora cardiaca* (Fig. 70).

Other parts of the nervous system include an *unpaired ventral nerve* that controls the spiracles and a *caudal sympathetic system.* The latter system arises from the last compound ganglion of the ventral nerve cord and sends nerves to the reproductive system and posterior part of the alimentary canal.

The nerves radiating from the ganglia, especially in the abdomen, to various structures are simple, when compared to vertebrates, and contain few neurons. Each neuron is able to transmit impulses of only one magnitude. Sensory information is coded into the system according to the number of impulses, their frequency, and the number of neurons involved. Low-level stimuli may only produce a simple reflex through the segmental ganglion. Strong stimuli, however, produce a great number of impulses along many sensory neurons which "overflow" at synapses into additional neurons and result in integrated motor patterns. In addition, two nerves run to each skeletal muscle, one a "slow fiber" and the other a "fast fiber." Suitable combinations of impulses between the two fiber tracts produce the necessary responses.

REPRODUCTIVE SYSTEM

One of the major problems of a dioecious species (one with male and female individuals) is getting sperm to the egg. In the water, sperm can simply be released near or on the ova, but the dry terrestrial environment does not normally permit this adaptation. Unprotected sperm and ova quickly dessicate.

Ova are protected by an exoskeleton or *chorion.* Pores, or *micropyles,* permit sperm penetration. Other modifications include differences in shape, various sculpturing in the chorion, and specialized pores, the *aeropyles,* that permit oxygen uptake yet restrict water loss. Cytoplasm of the egg is generally located peripherally, but some extends out through the yolk to the nucleus (Fig. 71B).

Insect sperm are unique in that the flagellum has only a short end piece. In the least specialized insects sperm are protected by gelatinous membranes in specialized sacs or *spermatophores.* These spermatophores are either deposited

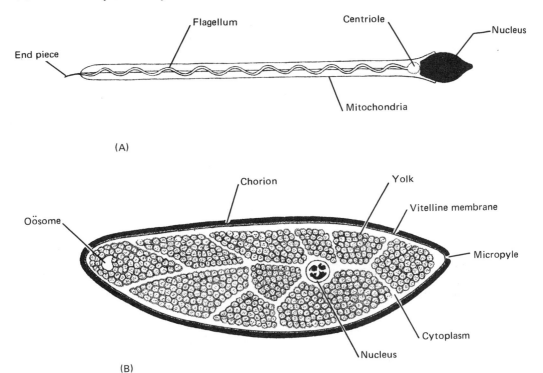

Figure 71.   Insect gametes. (A) sperm; (B) egg.

on the soil surface for the female to locate, as in some Collembola, or are transferred into the body of the female during copulation. Spermatophores have been lost in various insects or are represented only by vestiges. A summary of the evolution of sperm transfer is given in Figure 72.

Male

The gonads in this sex are two *testes* consisting of a definitive number of *sperm tubes* enclosed in a membranous sac. Within these tubes (Fig. 73) unspecialized cells, or *spermatogonia,* increase in size to become *primary spermatocytes,* undergo meiosis to become haploid *secondary spermatocytes* and *spermatids,* and are then transformed into *sperm.* Leading from each testis is a *vas deferens* (Fig. 74), which carries these gametes down to the single *ejaculatory duct.* The *accessory glands,* located either as outpouchings of the vas deferens and/or ejaculatory duct, add fluids and a covering if spermatophores are used. During copulation, sperm is transferred into the female genital system by the *aedeagus.*

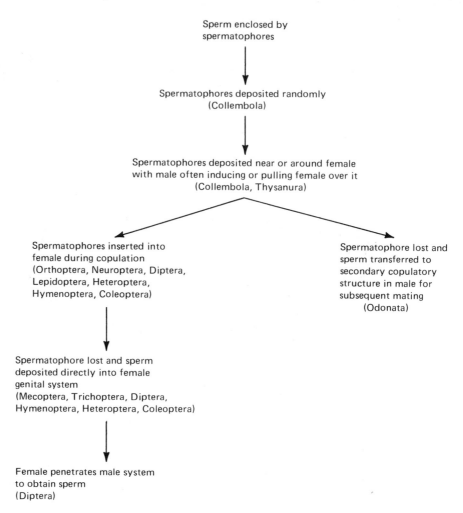

Sperm enclosed by
spermatophores

Spermatophores deposited randomly
(Collembola)

Spermatophores deposited near or around female
with male often inducing or pulling female over it
(Collembola, Thysanura)

Spermatophores inserted into
female during copulation
(Orthoptera, Neuroptera, Diptera,
Lepidoptera, Heteroptera,
Hymenoptera, Coleoptera)

Spermatophore lost and
sperm transferred to
secondary copulatory
structure in male for
subsequent mating
(Odonata)

Spermatophore lost and sperm
deposited directly into female
genital system
(Mecoptera, Trichoptera, Diptera,
Hymenoptera, Heteroptera, Coleoptera)

Female penetrates male system
to obtain sperm
(Diptera)

Figure 72. Evolution of sperm transfer in insects.

Female

The female system has a dual role, to receive the male aedeagus and to produce ova. During copulation, sperm is released either into the *genital chamber* (Fig. 74) or into a specialized diverticulum from the genital chamber, the *bursa*. These gametes migrate or are moved by peristalsis up into the *spermatheca* via the *spermathecal duct* where they are stored until an egg is released. The length of survival of sperm varies from only several weeks in bed bugs to many years in queen ants. In the latter situation food of some sort must be supplied and the metabolism of the sperm must be slowed down, but the actual mechanics remain uncertain.

    The gonads of this sex consist of two *ovaries,* which consist of a definite number of tubes or *ovarioles* (Fig. 75) linked apically by their collective *terminal filaments. Oögonia* are located in the terminal region and enlarge into *primary oöcytes* as they move down the ovariole and as they receive yolk from the follicular epithelium (*panoistic ovarioles*) or from special nurse or nutritive cells (*meroistic ovarioles*). This enlargement produces a distension of the

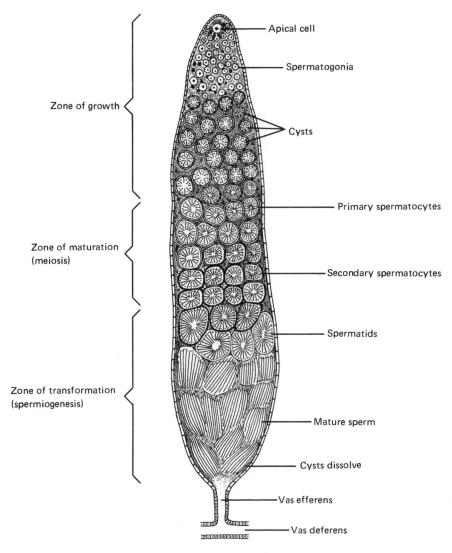

Figure 73. Diagram of sperm tube or follicle indicating sperm development. (Redrawn with slight modifications from Snodgrass, 1935)

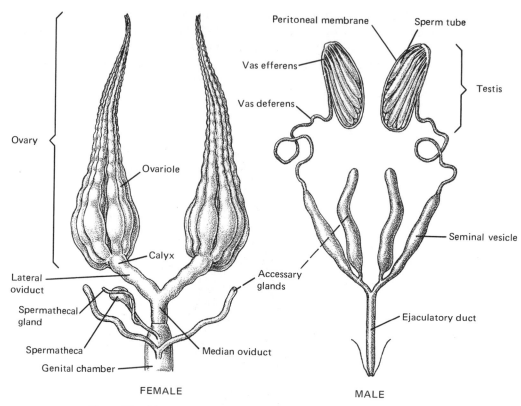

Figure 74. Diagrammatic representations of the female and male reproductive systems. Much variation from these hypothetical models occur. (Redrawn with slight modifications from Snodgrass, 1935)

ovariole into a *follicle*. Finally, a chorion is secreted that gives the characteristic species shape and covering.

After leaving the ovariole, eggs are often stored in the *calyx,* particularly in insects that deposit large numbers of eggs at a single laying. The egg, oöcyte at this stage, is then moved by peristalsis down the *lateral oviduct* and the *median oviduct* to the spermathecal duct pore where sperm are released and penetrate the micropyle. The sperm entry initiates oögenesis in which the oöcyte nucleus undergoes meiotic division to produce a haploid gamete nucleus or ovum and polar body nuclei. The polar body nuclei are lost, but the ovum and sperm nuclei fuse and the normal number or diploid state of chromosomes is restored. Once the sperm has penetrated, the egg is moved down the genital tract and is coated with various materials from the *accessory glands* (when present) for sticking the eggs to a specific substrate. *Oviposition* then occurs.

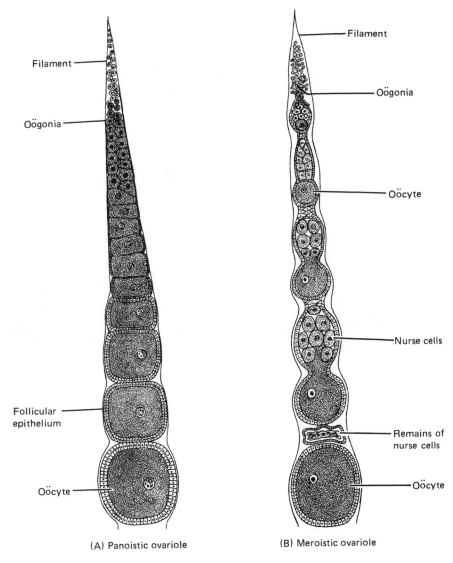

Figure 75. Two types of ovarioles. (A) panoistic from Orthoptera; (B) meroistic from Hymenoptera. (Redrawn with slight modifications from Imms, Richards, & Davies, 1957)

Sexual Reproduction

Oviparity. The majority of insects undergo the type of reproduction previously described in which an egg is produced, fertilized, and oviposited by the female. This is oviparity (Fig. 77). Intricate behavior patterns are involved and the eggs are normally deposited in precise microhabitats, near or on the required food, etc. Sufficient yolk is present to permit embryology to be completed within the egg.

81

Figure 76. An American female cockroach (*Periplaneta americana*) carrying an oötheca.

Eggs may be deposited singly, in groups, or "fused" together into an *oötheca* (Fig. 76) by accessory gland secretions, as in cockroaches. Mortality from water loss is slowed down by a waxy layer secreted beneath the chorion by the egg itself.

Ovoviviparity.   In ovoviviparity (Fig. 77), eggs are normally developed and fertilized, but they are retained and hatched within the body of the female. Sufficient yolk is present in the eggs for the embryo to complete development. The number of eggs produced may be restricted, but the protection offered plus the depositing of immatures which are ready to commence feeding are obviously of selective advantage. Flesh flies are good examples of this variation, and the larvae are deposited on fresh carcasses without immediate competition from other larval insects.

Viviparity.   In viviparity (Fig. 77), development takes place within the female body. In some variations the eggs do not contain sufficient yolk to permit the embryos to develop fully, and the young, underdeveloped embryos, hatch to be nourished by the mother. Others have little or no chorion and are nourished continuously. Aphids, during the summer months, and certain flies, such as the tsetse of Africa, are good examples of this type.

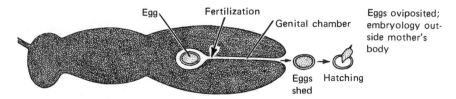

(A) Oviparous reproduction

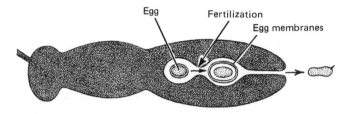

(B) Ovoviviparous reproduction

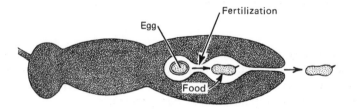

(C) Viviparous reproduction

Figure 77. Comparisons between oviparity, ovoviviparity, and viviparity.

Asexual Reproduction

Polyembryony. In this variation, the dividing cells, instead of cleaving together, separate during the initial mitotic divisions and each cell subsequently gives rise to a separate individual. The number arising from each egg varies from two to several thousand. Polyembryony is normally restricted to a few parasitic species and is beneficial because it permits a small parasite to exploit a large host and its food supply with minimal danger to the female as she oviposits. In a restricted sense it represents a type of asexual reproduction.

Parthenogenesis. Some insects are capable of reproducing without fertilization (Fig. 78). The resulting individuals may be diploid either because no meiosis occurred (aphids) or because a polar body fused with the egg to form the zygote (some walking sticks and lepidoptera). In other instances, sex is

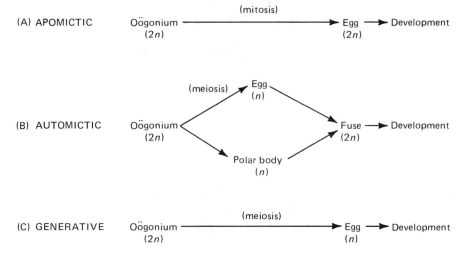

Figure 78. Types of parthenogenesis. Type A has the greatest genetic stability.
Type C determines sex as the opposite is diploid from normal fertilization.

determined by parthenogenesis with either the male (Hymenoptera) or females (various species) developing from nonfertilized eggs and hence being haploid. Parthenogenesis is normally associated with the other types of reproduction such as viviparity (aphids) and oviparity (bees).

A major disadvantage of parthenogenesis is the relative genetic stability from generation to generation. The offspring are usually the same genetically except for mutations that cannot be distributed as rapidly into the species gene pool as can occur in oviparity.

Paedogenesis (Neotony).   In a few instances, *immatures* are capable of producing offspring by parthenogenesis, often as part of an extremely complex life cycle involving alternation of generations. Unlike most insects, the development of functional gonads occurs prior to metamorphosis thereby permitting this rare phenomenon.

Questions

1.   How is the surface/volume ratio important in the development of body systems of an organism?

2.   What is digestion? How does a knowledge of the embryonic origin of the insect alimentary canal assist one in understanding what occurs in each portion?

3.   Compare a digestive tract specialized for solid food with one

modified for a liquid diet. Does the amount of water in the food have an effect upon digestion and the modifications of the system?

4.   Correlate liquid or solid food types of alimentary canals with the types of mouthparts discussed in Chapter 2.

5.   How might symbionts be important in insect nutrition?

6.   Characterize the insect circulatory system and the directional flow of hemolymph throughout the body. Does the size of the insect have any effect upon the degree of specialization?

7.   How does the circulatory system of an insect differ from that of a human?

8.   What is ATP? Where does the energy necessary to synthesize ATP come from? What is its function?

9.   What is the importance of tracheoles? Discuss the role of tracheal trunks and air sacs.

10.   What modifications of the tracheal system are often seen in aquatic insects? What advantages are gained from these?

11.   How does the size of an organism affect the muscular system?

12.   What are the advantages and disadvantages of asynchronous muscles?

13.   How does flight occur in insects? How important is the notum or tergum to this phenomenon?

14.   What advantage is gained by depositing excretory materials into the feces for evacuation? Is there a difference in excretion between insects that eat solid foods and those that ingest liquids?

15.   What is a ganglion? What role does it play in the nervous system?

16.   How does terrestrial existence influence reproduction?

17.   What is the significance of meiosis? Is the egg that leaves the ovary a gamete?

18.   Compare the types of reproduction. What are the advantages and disadvantages of each?

4

Development and Specialization

EMBRYOLOGY

In most insects, penetration of the egg by a sperm initiates meiotic divisions resulting in a haploid nucleus that fuses with the sperm nucleus. The zygote nucleus next undergoes a series of mitotic divisions. Some of these nuclei remain behind to become *vitellophags,* cells for metabolizing yolk for embryonic use, but most of the cleavage nuclei migrate out through the radiating cytoplasm to the periphery of the egg (Fig. 79). Here, each produces a cell membrane, and the peripheral cells form the *blastoderm.* Some of this blastoderm will become part of the embryonic coverings (serosa and amnion), and a thickened area or *ventral plate* will give rise to the embryo (Fig. 79). During the next stages, the developing embryo may migrate through the yolk, a process called *blastokinesis.* These movements are characteristic of eggs that have much yolk, as is common in insects undergoing incomplete metamorphosis, but eggs that have little yolk have reduced movements or the process is absent.

Johannsen and Butt (1941) and Hagan (1951) give detailed information on the development of embryos. In summary, the ventral plate forms a double layer of cells, the presumptive *ectoderm* and *mesoderm.* Major structures derived from these layers are diagrammed in Figure 80. The presence or absence of a third layer, the endoderm of most animals, is inconclusive. The ventral side, including the nervous system, develops first; then the digestive

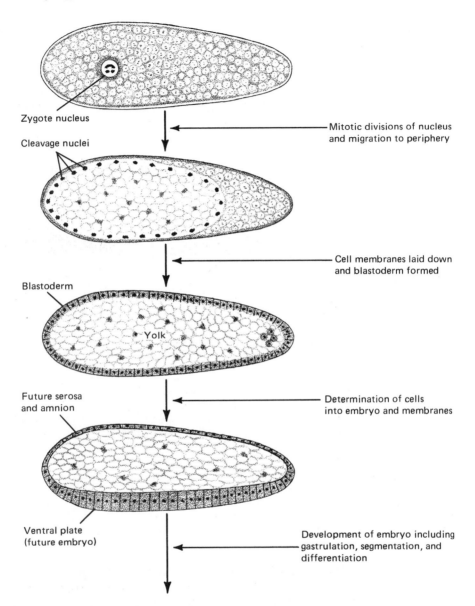

Zygote nucleus

Cleavage nuclei

Mitotic divisions of nucleus
and migration to periphery

Cell membranes laid down
and blastoderm formed

Blastoderm

Yolk

Determination of cells
into embryo and membranes

Future serosa
and amnion

Ventral plate
(future embryo)

Development of embryo including
gastrulation, segmentation, and
differentiation

Figure 79.   Early stages within the insect egg.

system; and lastly a phenomenon called *dorsal closure* produces the sides and dorsum of the embryo including the dorsal vessel. A cavity between the alimentary canal and exoskeleton, the hemocoel, develops, but it is not the coelom as in Annelida. Segmentation and tagmosis in the exoskeleton, muscles, and nervous system occur early although complete partitions between the segments of metameres never develop (Fig. 81).

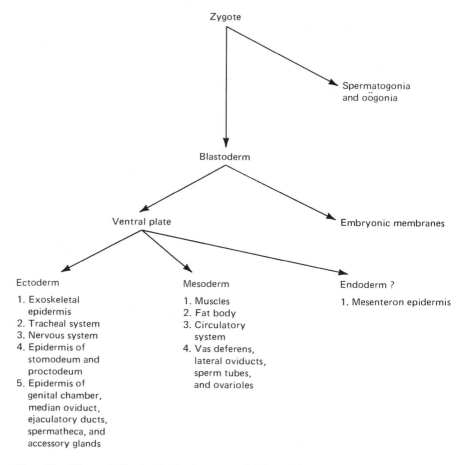

Figure 80. Diagram indicating the development and differentiation from the zygote to the organism with special reference to organ derivation.

The rate at which development occurs varies mainly with (1) temperature since insects are cold-blooded and (2) diapause or suspended development induced by the environment or obligatory by genetic control. The latter are especially pronounced where cold winters or long period of drought occur with regularity.

Emergence from the egg is accomplished in various ways. The immature may either chew its way out or digest a portion of the chorion. In other species the insect pushes against a preformed weakened area in the egg or uses specialized hatching spines or bursters. The stimulus or stimuli that initiates emergence is poorly understood except in a few species.

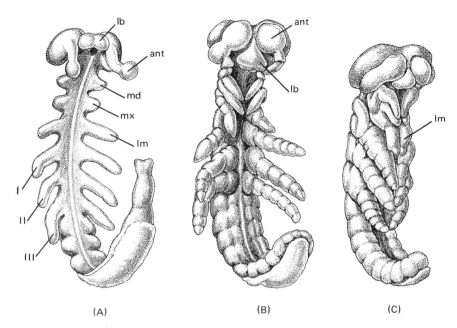

Figure 81. Embryology of the milkweed bug, *Oncopeltus fasciatus.* Envelopes and yolk have been dissected away. (A) 71 hours; (B) 91 hours; (C) 101 hours. ant, antenna; md, mandible; mx, maxilla; lb, labrum; lm, labium; Roman numerals, thoracic legs. (Redrawn after F. H. Butt, 1949.)

POST-EMBRYONIC DEVELOPMENT

Growth and Ecdysis

Growth in most animals represents the synthesis of new protoplasm and is normally accompanied by cell division. In arthropods this process is limited by the size and ability of the exoskeleton to enlarge. Once maximal size is reached, the restrictive integument must be removed and replaced by a larger one if further growth is to occur. How is this possible? If the new one is laid down under the old one, surely the insect would get progressively smaller with each *molt,* or *ecdysis.* If deposited externally, the old one would have to be somehow digested or split. Molting is as follows. First, the epidermis loosens from the cuticle and a new integument is secreted under the old one (Fig. 82) with much of the building blocks or metabolites coming from digesting parts of the original skeleton (up to 85 percent may be digested and absorbed). Initially, the new integument is unsclerotized and somewhat folded. The insect then draws in air and contracts its muscles, thereby increasing its body pressure until the old exoskeleton breaks along certain preformed weak points,

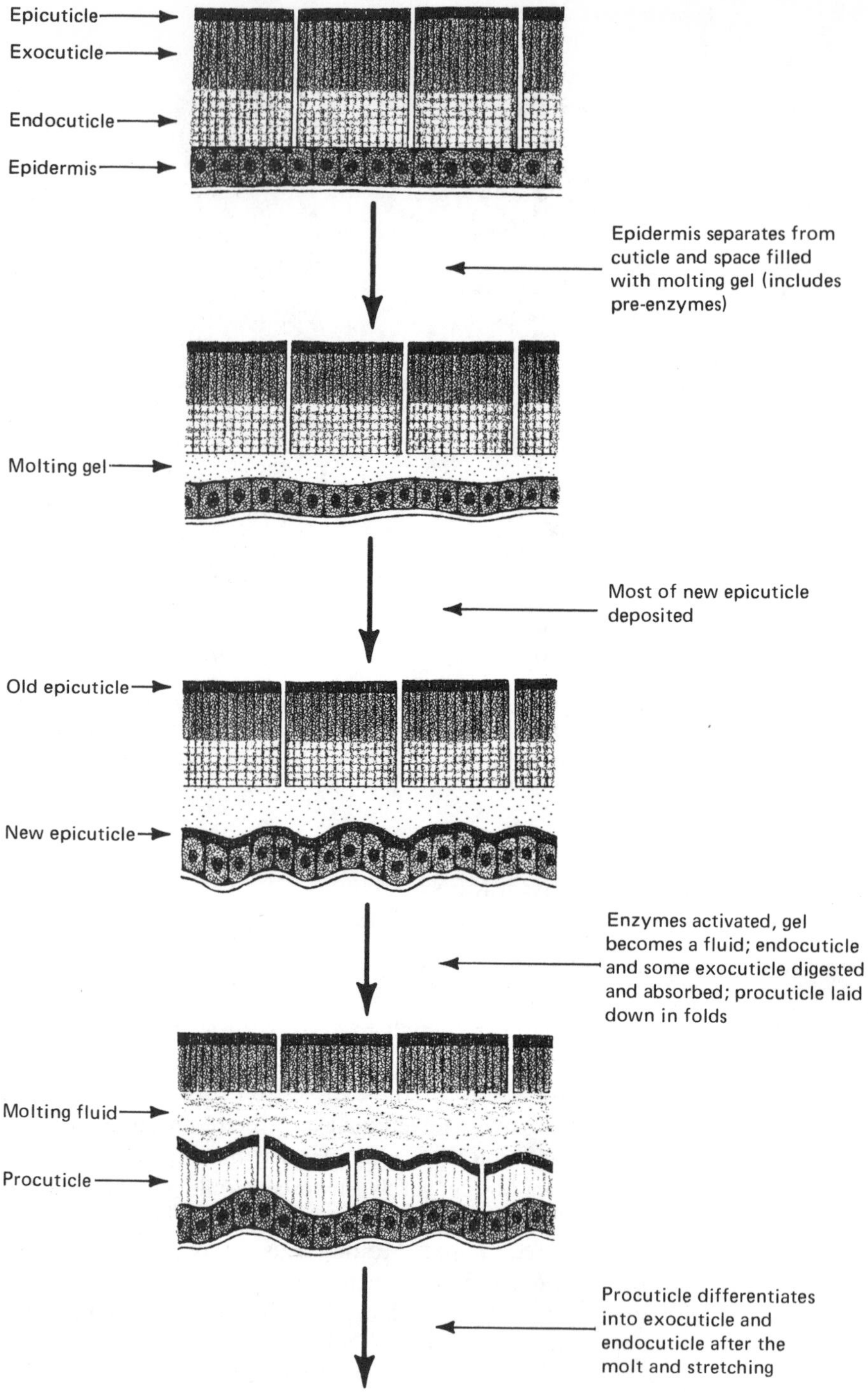
Epicuticle
Exocuticle
Endocuticle
Epidermis
Epidermis separates from cuticle and space filled with molting gel (includes pre-enzymes)
Molting gel
Most of new epicuticle deposited
Old epicuticle
New epicuticle
Enzymes activated, gel becomes a fluid; endocuticle and some exocuticle digested and absorbed; procuticle laid down in folds
Molting fluid
Procuticle
Procuticle differentiates into exocuticle and endocuticle after the molt and stretching

Figure 82. Sequence of events in the molting of the insect exoskeleton.

or *ecdysial sutures,* in the head and thorax. By peristaltic body movements, the old skeleton, or *exuvium,* is moved off posteriorly (Figs. 83, 87) with any remaining molting fluid acting as a lubricant. Because high body pressure is maintained, the new integument is straightened and stretched. Once the integument is expanded, tanning or sclerotization occurs and the remaining layers of the epicuticle are deposited. After the exoskeleton hardens, the muscles relax, air is exhaled, and space is now available for further protoplasmic synthesis.

All parts of the body lined with a cuticle take part in the molt, including the integumental linings of the trachea, stomodeal, and proctodeal intima. It should be obvious from the complexity of this phenomenon that ecdysis is an extremely critical time in the organism's life. The protection once afforded by the exoskeleton is removed and the insect is highly vulnerable to predation, chemical and physical forces, and water loss. Water loss is often minimized by limiting the process to periods of high humidity; nevertheless, water loss is critical (up to six times normal) and the molt must be completed rapidly. Occasionally an insect can be found that has parts of the old exoskeleton still adher-

Figure 83. Cockroach *(Leucophaea maderae)* in the process of molting. Note the ecdysial split in the thorax and head which permits the old exoskeleton to be shed.

Figure 84. Instars of cockroach (*Leucophaea maderae*). The top row are fully sclerotized individuals while the bottom row are newly molted.

ing to its body (Fig. 111); this indicates a failure of the ecdysial suture to split or too rapid evaporation of the lubricating molting fluid. If vital areas such as the tracheal system or head capsule, for example, are not shed properly, the organism invariably dies.

Although most protoplasm is added between molts, ecdysis superficially gives growth a discontinuous appearance and divides the individual developmental cycle into a series of steps or *instars* (Fig. 84). An instar is best defined as the individual between molts. The first instar represents the insect from hatching, or *eclosion,* to the first molt. Although variation does exist, the number of molts and instars for a species is relatively constant. The average insect completes approximately from four to six. Deviations from the species norm can result from nutritional deficiencies, temperature, and the sex of the individual.

Injuries to the integument must be repaired prior to a critical point for molting to proceed. Mending involves an aggregation of cells across the wound followed by the deposition of a new localized cuticle that extends under the old one at the wound margin for a short distance. Regeneration of lost or injured parts is also possible in some species, but these appear only during a molt.

Metamorphosis

In the most primitive insects, such as Collembola and Thysanura, body proportions and internal organs remain similar after each molt or ecdysis.

Species undergoing such a development are categorized as *Ametabola,* indicating no change in metabolism. Less than 1 percent of all insect species exhibit this growth pattern. The immatures live in the same habitat and feed on the same food as adults, and both immatures and adults are capable of molting.

Most insects, however, undergo changes in shape or form during their development, a process of *metamorphosis.* Some species have only minor modifications, but a gradation may be seen up to very drastic alterations, e.g., from legless wormlike immature individuals to adults with legs and wings as seen in flies. Perhaps no concept in entomology has resulted in more polarization of thought than that of categorizing metamorphosis into types and explaining how such differences evolved. One classification includes insects that have immatures with external wing pads, the *exopterygota,* and species that have interal wing pads as larvae, the *endopterygota.* Another classification agrees with the previous demarcations but substitutes *hemimetabola* for exopterygota and *holometabola* for endopterygota. Still another view separates the exopterygota into *paurometabola* (for those whose immatures develop terrestrially) or *hemimetabola* (whose immatures inhabit water) and retains the endopterygota as a single unit, the holometabola. Fully realizing that no system is completely adequate, this text will use the terms *hemimetabola* and *holometabola,* because of their more worldwide usage, to illustrate development and metamorphosis.

When a change from the immature to the adult occurs but is not extreme, metamorphosis is *incomplete* or hemimetabolous. Immatures or *nymphs* (called larvae by some) are normally characterized by possessing external wing pads (Figs. 42, 85, 86), except when wings are secondarily absent as in lice. Most structures of nymphs and adults (Fig. 90) with incomplete metamorphosis are alike although body proportions differ and changes in the thoracic plates and reproductive system occur in the molt to an adult. Food and ecology are similar in all stages. The greatest disparity in the *hemimetabola* exists in aquatic immatures, sometimes called *naiads,* in which some specialized structures are vital for survival in water. Odonata nymphs have an enlarged labium (Fig. 108) for capturing prey and gills of two types, the three flat taillike tracheal gills of damselflies and the internal rectal gills of dragonflies. Mayfly nymphs have a pair of gills (Fig. 231) for most abdominal segments, and many species have either or both enlarged tusklike mandibles and the front pair of legs for digging. These immature characters, however, are lost in the metamorphosis to adulthood.

The majority of insect species, approximately 85 percent, have a *pupal stage* (Fig. 85) and undergo more drastic alterations. Metamorphosis is *complete* (holometabolous development), and often the majority of the larval structures are broken down by histolysis, and adult structures within the immatures are built from either small groups of adult tissues, the *imaginal discs,*

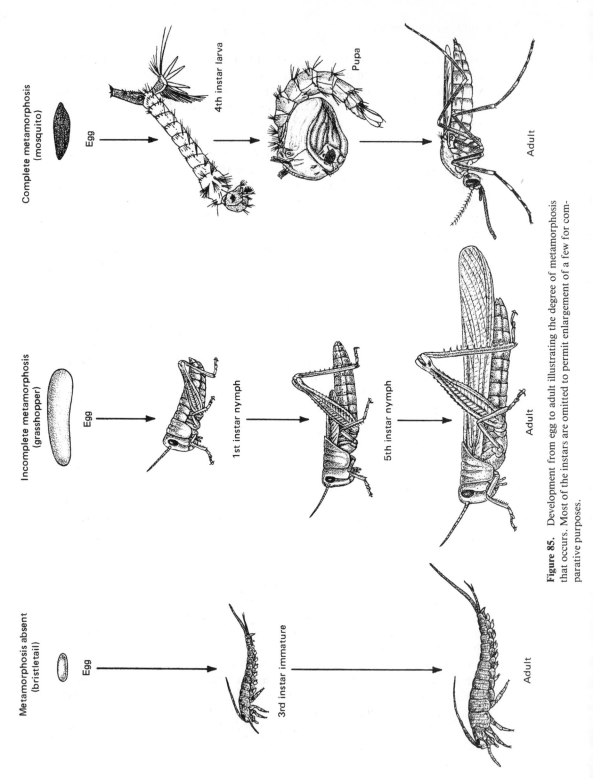

Figure 85. Development from egg to adult illustrating the degree of metamorphosis that occurs. Most of the instars are omitted to permit enlargement of a few for comparative purposes.

Complete metamorphosis (mosquito)

Egg

4th instar larva

Pupa

Adult

Incomplete metamorphosis (grasshopper)

Egg

1st instar nymph

5th instar nymph

Adult

Metamorphosis absent (bristletail)

Egg

3rd instar immature

Adult

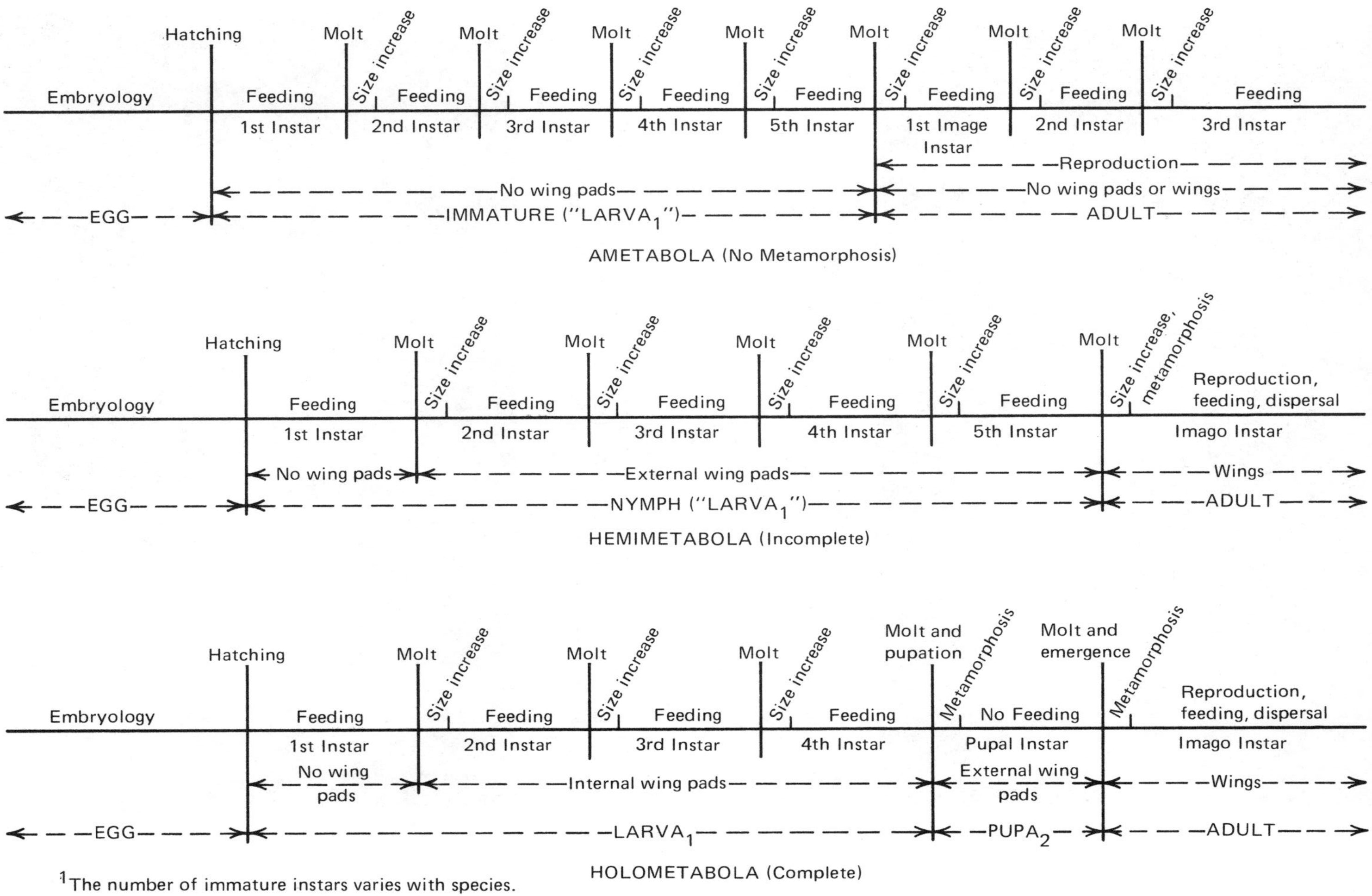

Figure 86. Major types of development in insects. (*After* Fernald & Shepard, 1942.)

[1] The number of immature instars varies with species.

[2] The pupa is considered the 1st imago or adult instar by some.

95

Figure 87. *Hyalophora cecropia* larva molting. The old exoskeleton is thin and has been capable of limited stretching prior to the molt. The white bands along the side represent tracheae being shed, a process necessary since they are lined with cuticle. Note the size difference between the old head capsule and the new one.

Figure 88. A *Hyalophora cecropia* larva transforming into a pupa within a protective cocoon consisting of three layers, the inner smooth to protect from injury during molt, the middle loosely arranged for insulation, and the outer for proper background color and water repellency. Note the change in antennal shape, type of legs, as well as the tracheae being molted with the larval exoskeleton.

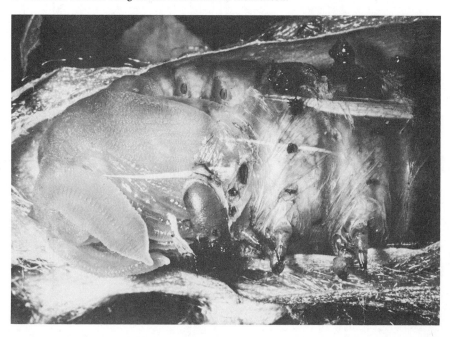

or from larval cells. Wing pads are internal in larvae (Fig. 87) but evert to become external in the pupa (Fig. 88) and enlarge to functional wings in the adults. This type of reconstructive development is found in most advanced species. Larvae tend toward becoming the dominant feeding stage, the pupa becomes essentially a transformation stage, and the adult is specialized for dispersal and reproduction. Although most adults do feed, some survive only upon nutrients stored in a fat body built up during the larval stage. In many species, this apparent division of biological labor has enabled extreme exploitation of an environment, especially where resources are seasonally low, and reduces intraspecific competition between young and adults.

The pupal stage is of three types. *Exarate* pupae (Fig. 109) do not have their appendages appressed to the body, a situation found in Neuroptera, Trichoptera, most Coleoptera, some Lepidoptera, and most Hymenoptera. A second type, the *obtect* (Fig. 89), have their appendages fused to their body, a state common in Lepidoptera, some Coleoptera, and the lower Diptera. The third type are enclosed within the last larval exoskeleton, the *puparium* (Fig. 91), and are classified as *coarctate*. Pupae, in general, exhibit only limited

Figure 89.   After completing diapause, the *Hyalophora cecropia* pupa completes the development to an adult. Note the adult color patterns under the pupal exoskeleton as the final molt is about to occur.

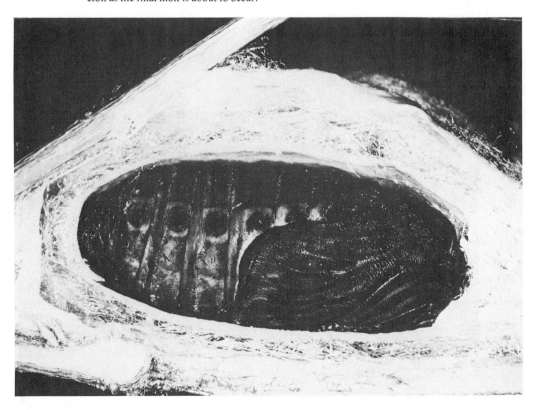

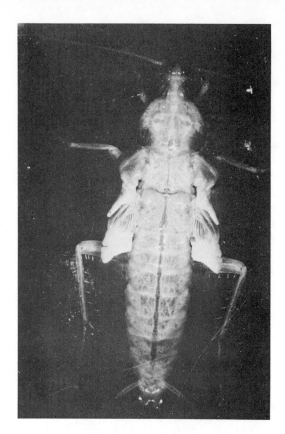

Figure 90. A freshly molted adult American cockroach (*Periplaneta americana*) in the process of expanding its wings. Once expanded, the wings become sclerotized and the flight muscles mature prior to adult activity. Note the heart and tracheae showing through the unsclerotized integument.

Figure 91. Flesh fly *(Sarcophaga bullata)* adult emerging from the puparium, a protective sac consisting of the exoskeleton of the last larval instar. Note the ptilinum, an eversible structure at the front of the head which is used to escape from the puparium.

98

movement and most pupae are sessile and found in protected environments such as in the ground, under bark, in cases or retreats fashioned by the larvae, or in silken cocoons produced by the last instar larvae.

Both growth and metamorphosis are affected by temperature since insects are cold-blooded animals. At 30 °C, a house fly becomes a pupa 5 ½ days after hatching from an egg, but at 10 °C it requires 34 days. Similar examples may be seen in nearly all species.

Diapause

During embryonic development, growth in the post-embryonic stages can be halted resulting in a dormancy, or *diapause.* This phenomenon permits an insect to survive under seasonal situations when normal metabolism would produce death through lack of food, exhaustion of stored fat, or freezing. Most temperate species and some tropical ones enter into diapause during some stage in the life cycle. In many instances, this dormancy is genetically programmed and reversal occurs with difficulty only after unfavorable conditions have passed. The photoperiod of increasingly shorter days, often correlated with cool days, normally induces diapause in temperate regions whereas the dry season triggers it in the tropics.

There are two basic types of diapause, *obligatory* and *facultative.* The former always occurs at the same stage and is common in univoltine species, those with only one generation each year (Fig. 92). When diapause is obligatory, the environment affects only the duration of diapause and not when it commences. Facultative diapause occurs only when environomental conditions dictate and is probably the most common type. Such stimuli as lowered moisture, temperature, or food often induce facultative diapause.

Diapause is known from every life stage. In the mosquito genus *Aedes,* the embryo develops to a larva prior to entering dormancy. Hatching occurs only after the egg is submerged and the oxygen pressure is lowered. These mosquitos are able to survive during freezing conditions or in the equally xeric conditions of the desert. In another example, the *H. cecropia* moth (Fig. 89), diapause occurs in the pupa and is broken only after an extended period of freezing.

Regeneration

Insects are capable of limited amounts of regeneration, depending on the species, age of the individual, and the extent of injury. In contrast to vertebrates and annelids in which the process is similar to the original embryology, regeneration in insects starts with a division and migration of epidermal cells around the lost structure. Nerves and trachae then grown into the regenerate area. Muscle formation hasn't been determined although this tissue

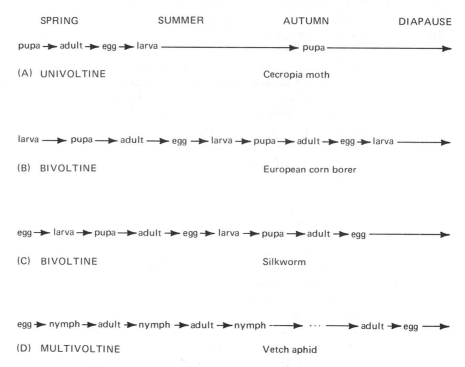

Figure 92. Insect species illustrating, (1) stages in which diapause occurs, and (2) the number of generations per year that each passes through prior to overwintering: Univoltine = 1 generation per year; Bivoltine = 2 per year; and Multivoltine = many per year.

does require the presence of nerves in order to develop (Nuesh, 1968; Goss, 1969). With subsequent molts, part or all the lost structure will develop, depending on the species and the extent of the earlier loss.

Some insects such as walking sticks have weak points in their appendages and may lose these structures when they are grasped by a predator. The loss affords the insects a means of escape. The limb often twitches and diverts attention away from the escaping host. Such a phenomenon is termed *autotomy*. Subsequent molts result in replacement although the legs may be reduced in size.

Insects that undergo complete metamorphosis are unique in that the loss of larval appendages, unless very extensive, usually does not appear in the adult. This is because most adult structures arise from imaginal discs that develop into the adult structures at metamorphosis.

ENDOCRINE SYSTEM

An endocrine structure is a gland or tissue that produces *hormones* (chemical messengers) and either secretes them into the blood or into the external environment. In insects these "glands" are often more appropriately referred to as *tissues* because of their relative simplicity. They very commonly are nervous tissues that have been modified for hormonal production and secretion.

Hormones are characterized by their ability to produce specific reactions at minute quantities upon specific tissues (*target structures*). The reaction can be stimulatory or inhibitory; not all tissues are equally competent or ready to respond to the presence of hormones because the cellular environment in various portions of the body is continually fluctuating.

The presence and effect of hormones are difficult to demonstrate. Once the presence of a hormone is suspected (often suggested accidentally), scientific evidence of cause and effect must be determined and usually involves the removal of a tissue with subsequent implantation of the tissue of extract in the same or different individual. Because of the small size of insects and the difficulty in isolating tissues and compounds, until recent years there was much doubt that hormones existed.

Growth and Metamorphic Hormones

The first evidence of hormones and the most widely studied ones today are those involved in the phenomena of growth and metamorphosis. The discovery and techniques involved in elucidating their presence and role provide some of the most fascinating experiments in biology. Readers are referred to such works as Williams (1947, 1952, 1958) and Wigglesworth, (1954, 1970). One experiment (Williams, 1952) is summarized in Figure 93.

Certain neurosecretory cells in the protocerebrum produce the *brain* (= activation or prothoracictropic) *hormone.* The number of cells involved in producing this hormone varies from approximately five to several thousand, depending on the insect studied. Once synthesized, the hormones are moved along the axons of these neural cells from the brain to the *corpora cardiaca* (Fig. 70) where they are released into the hemolymph at specific times. The target tissues are the *prothoracic glands* which become activated.

The prothoracic (= thoracic or ecdysial) "glands" are inconspicuous tissues often associated with the first pair of thoracic spiracles. Although small, their effect is monumental, for they produce a group of hormones, the *ecdysones* (= growth and differentiation or molting) that activate the epidermal cells to produce molting fluid. When an inhibitory hormone is absent or in low concentrations, ecdysones also induce metamorphosis of the immature to an adult.

A third group of hormones, often referred to singularly as the *juvenile hormone,* are produced in the corpora allata (Fig. 70) and maintain the proper functioning of larval genes. When both ecdysones and juvenile hormones are present, growth and molting occur, but the larval characteristics are perpetuated. In some genetically predetermined instar, juvenile hormone synthesis is either shut off or the level drops to a low level that results in the inac-

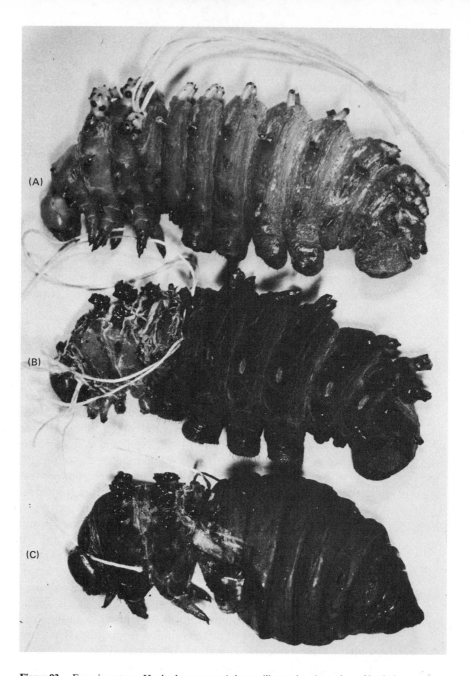

Figure 93. Experiments on *Hyalophora cecropia* larvae illustrating the action of brain hormone and ecdysone by ligation. The juvenile hormone is no longer being produced. The procedures followed were similar to Williams (1952). Ligation by body regions prevented distribution of hormones except within certain areas. The thorax region in B and C were not able to shed the larval exoskeleton because of the ligations, and the head, deprived of food and oxygen, died. (A) ligation before any hormones were released. No metamorphosis or molting took place; (B) ligation after brain hormone had been released but prior to release of ecdysone from prothoracic glands. Only the thorax metamorphosed, the head and abdomen remained in their larval form; (C) ligation after both brain hormone and ecdysone were released. All regions underwent metamorphosis.

tivation of these larval genes and the subsequent differentiation of adult characteristics through the action of ecdysone. If the immature is similar to the adult (incomplete metamorphosis), the adult will be produced at the next molt. If, however, drastic alterations must be made (complete metamorphosis), then two molts are required; the first instar is referred to as the *pupa* and the second instar is referred to as the *adult*. Compounds that mimic the juvenile hormone have been found in plants.

Interesting experiments may be imposed upon this system. Removal of the corpora allata in early instars may result in premature metamorphosis and a dwarf adult. In contrast, implantation of corpora allata or injection of juvenile hormone into a last instar may produce an additional instar and a giant individual (Fig. 94).

Pheromones

Pheromones may be defined as hormones secreted externally that affect other individuals of the same species. Only minute concentrations are required for biological activity. Some pheromones have an immediate effect on the central nervous system and behavior, "releaser substances," while others trigger a chain of developmental events, "primer substances" (Wilson 1963).

Sex Pheromones.   Over a hundred years ago, J. H. Fabre observed that upon placing a female moth in a cage outside his house, male moths of the same species were attracted in large numbers from nearby woods. No females were drawn in. Fabre speculated that, because of their disproportionate size, the large male antennae were detecting something that was attractive. Little follow-up on this and similar observations was made until recently when advanced technology provided a means of isolating, analyzing, and testing compounds that appear to induce these responses. Research is providing a picture of much diversity in molecular chemistry and action of these attractants, but two generalities may be made. First, it is normally the female that releases the pheromones and the male that is attracted. Second, sex pheromones are usually produced in abdominal glands although a notable exception is in the mandibular glands of the queen honeybee. Responses to these pheromones will be covered later in the chapter on behavior.

Trail Pheromones.   Trail pheromones are most common in ants and termites. As workers forage and return to the nest, periodic deposits are left by each individual. In some species, such as the fire ant, these pheromones are deposited by periodic dragging of the sting. In others, the substances originate from tarsal glands, abdominal glands, or from the alimentary canal. In most instances, since trail pheromones are highly volatile, continued use and pheromone deposition are necessary in order to maintain the trail. Consequently, it is only the routes from the nest to a good source of food that become trails (Fig. 95).

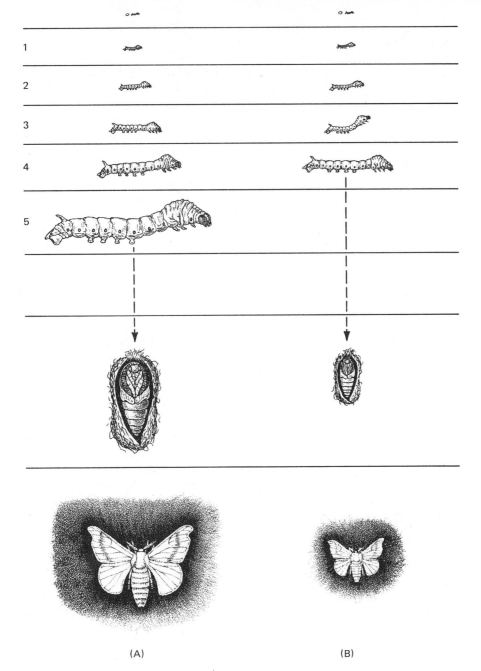

Figure 94. Dwarf and giant moths were made by removing and implanting the corpora allata. The column at left outlines the normal development of the commercial silkworm, *Bombyx mori.* At the top of the column is the egg of the insect and its newly hatched larva. The second column shows what happens when the corpora allata are removed at the 4th instar; the larva immediately changes into a dwarf pupa, and then into a dwarf moth. In the third column the corpora allata are removed at the 3rd instar, resulting in even smaller pupa and moth. In the fourth column the corpora allata of a young larva are implanted in a 5th instar larva. This larva continues to grow and then changes into a giant pupa and moth. (*From* "The Juvenile Hormone," C. M. Williams, 1958, Scientific American, Inc. All rights reserved.)

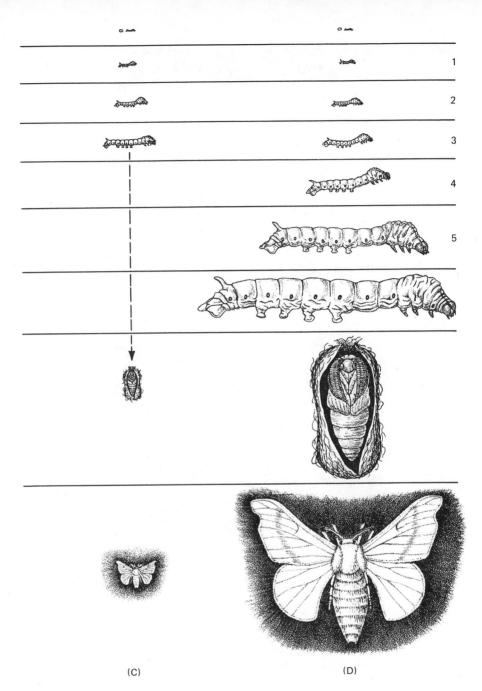

1

2

3

4

5

(C)

(D)

Figure 95. Leafcutter ants (*Atta* sp.) following a chemical trail across a piece of string back to the colony.

Figure 96. Termite nasutes (*Nasutitermes* sp.) attracted by alarm pheromones to a hole in the nest.

Since trail pheromones are detected by the antennae, the lower portions of these structures of trail-following species, such as army ants, have a disproportional number of receptors (Fig. 39).

Alarm (= alerting) *Pheromones.* Ants and termites produce pheromones, in addition to other substances, that excite their siblings to activity. The emergency that initiates the release may be a break in the nest wall, as in termites with above ground nests (Fig. 96), or an attack by predators. As additional individuals arrive on the scene, they too release these highly volatile alarm materials until the necessary number of individuals present are able to cope with the problem. Soldiers, when present, are especially activated. Different glands are responsible for the alarm secretion. In some ants, it is the mandibular gland; but in other ants, anal glands produce these substances. In termites, the frontal glands are suspected. The honeybee produces the alarm pheromone in the sting glands.

Additional Social Pheromones. Colony cohesion is often maintained by pheromones. A queen pheromone in honeybees and some social wasps inhibits maturation of the ovaries of workers, and should the queen die or leave, behavior changes drastically within the colony and a new queen is reared. Castes in termites are the result of pheromones produced by reproductives and soldiers. More details on these primer substances can be found in Chapter 7.

Other Hormones

Bursicon, a hormone produced by neurosecretory cells in ganglia, is necessary for darkening and sclerotizing the exoskeleton. In cockroaches, this hormone first appears after the old cuticle splits, but in other insects such as Lepidoptera and Diptera, its release may be delayed until after the adult escapes from either its cocoon or puparium. Bursicon production ceases several hours after molting.

Knowledge of hormones other than those directly involved with growth and metamorphosis is somewhat scanty. Production of juvenile hormones occurs after adult emergence from the pupa and it induces oöcytes to develop (= *gonadotropic hormone*). A *diapause hormone,* originating from the subesophageal ganglion, causes second-generation eggs of the silkworm moth to enter diapause. Undoubtedly, hormones are involved in certain color changes, water balance, and the varied responses of insects to light cycles (photoperiod), but scientific evidence is only suggestive. Most hypotheses include the neurosecretory tissues of the subesophageal ganglion, brain, and corpora allata as production sites of these additional hormones.

Questions

1. What factors influence embryology in insects?
2. Which cells in the developing egg do not become part of the embryo? What are the functions of these cells?
3. Why does growth in insects appear to be discontinuous or in sudden spurts? How does the new exoskeleton form?
4. Are growth and metamorphosis synonymous? Explain your position.
5. What are some of the advantages and disadvantages of complete metamorphosis (holometabolous development)? What is a pupa?
6. Is metamorphosis restricted to the exoskeleton and its resulting body shape? Explain your answer.
7. What hormones are involved in ecdysis? In metamorphosis? What is the function of each?
8. What are pheromones? Discuss their role in the life of the insect.

5
Ecology

The next logical step in unfolding the life of insects would have to be a discussion of the environment. Many current ecological studies deal with the *ecosystem,* or the study of the biotic and abiotic factors of an area and how they affect organisms (Fig. 97). Abiotic factors include temperature, soils, water, gases, to name a few, and are often the first to be considered. Biotic factors are organisms and their interactions within and between *populations,* localized breeding groups. Associations of many interdependent populations of different species constitute *communities;* groups of communities having similar dominant plant life forms make up a *biome.*

Because of the complexity of the natural environment, discussions will be restricted to a few of the major factors and concepts. For the student desiring more in-depth materials, texts such as Andrewartha (1971), Odum (1971), Boughey (1972), and Ricklefs (1973) are useful.

AQUATIC ENVIRONMENT

Although water may freeze or evaporate, it is usually found in a liquid state. Six characteristics of this substance must be discussed so that an existence of aquatic organisms can be understood. First, water gets heavier as it cools, until its maximal density is reached at 4°C. From this point to 0°C, it

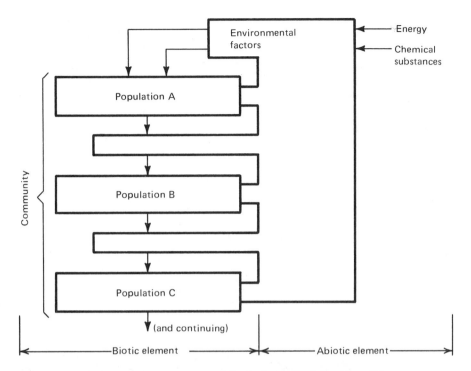

Figure 97. Basic features of an ecosystem and the fundamental relationships within it. Ecosystems are composed of biotic and abiotic elements. The biotic element is formed from groups of similar and related organisms, the *populations*. These are linked together by functional interdependence to form *communities*. The abiotic element is involved in processes of energy absorption and release of chemical substances. Both of these vital activities of an ecosystem are subject to environmental influences, as are the intracommunity and intercommunity biological interactions. (*From* A. S. Boughey, *Fundamental Ecology,* 1972.)

becomes lighter and floats to the surface before freezing. As a result, formation of ice progresses away from the surface, and whether or not a body of water freezes solidly will be determined by its depth and duration of subfreezing temperatures. Without this property, aquatic biota would not survive freezing weather.

Second, water can absorb large amounts of heat, almost 500 times that of air, without changing its temperature. This capacity to store energy produces a relatively uniform situation, when compared to air, because extremes in temperature within this environment are greatly moderated.

A third characteristic of water is its relationship to gases. The warmer the water, the lower the amount of oxygen that can remain dissolved. Warm water can be a limiting factor to life because of lowered oxygen levels long before temperature, itself, causes death. The correlation between temperature and the saturation levels of oxygen may be seen in Figure 102.

Fourth, water, in its liquid state, does not expand to fill space as gases do. Molecules that contact the atmosphere arrange themselves to form a distinct film, a membranelike feature caused by molecular cohesion of these surface water molecules. Some insects such as water striders (Fig. 98) and adult whirligig beetles use the surface film to move about on. Although most structures of water striders and whirligig beetles are similar to terrestial insects, the second and third pairs of legs are significantly modified. In water striders, these legs are greatly lengthened with the middle pair providing oarlike power strokes for locomotion and the hind pair being used for steering (Fig. 98). The tarsi of both pairs of legs are hydrophobic and prevent the insect from sinking. Whirligig beetles have their second and third pairs of legs shortened and flat for rapid swimming movements. In addition, the compound eyes of whirligig beetles are divided into dorsal and ventral halves (Fig. 99) with, apparently, the ventral portion functioning below the surface and the dorsal part for seeing in the atmosphere.

The fifth characteristic of water is the uneven distribution of electrons between oxygen and hydrogen, producing a polar molecule, i.e., its charges

Figure 98. The water strider (*Gerris remigis*) rows its way across the water using its middle legs for the power stroke. Note that the weight of the strider depresses the surface film.

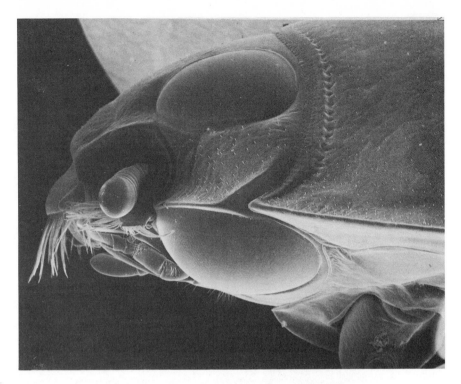

Figure 99. The adult whirligig beetle (*Dineutus americanus*) swims on the water surface. The eyes are separated into upper and lower sections to view above and below the water surface.

are not distributed symmetrically but form positive and negative poles. As a result, water molecules arrange themselves into a distinctive pattern that tends to neutralize any weak electrical fields around them. Substances held together by ionic bonding readily disassociate and dissolve. Also, the above hydrogen bonding pattern increases the ability of water to serve as a solvent for many other molecules. As a result, water varies in salinity and acidity. Obviously, physiological adaptations to survive under each pH situation must be present.

Last, water is dense and viscous. To be able to move through water, especially in currents, requires some form of streamlining. The entire body may be streamlined, as is usually found in insects that are strong swimmers, or only the dorsum may be rounded while the ventral surface may be flattened. For many years it was believed that streamlining of half the body caused the water passing over the body to press the organism tightly to the substrate, and, indeed, it does in a few instances (Fig. 100). In most insects, however, flattening is an adaptation that permits the insects to crawl into protected habitats to escape currents except for certain periods of feeding (Fig. 101). Avoiding direct exposure is certainly an efficient means of maintaining position in rapidly moving water.

Figure 100. Ventral view of a water-penny beetle larva (*Psephenus herricki*) found in rapidly moving water. Currents passing dorsally over the body press the organism tightly to the rock substrate. Note the abdominal gills.

Figure 101. Some caddisfly larvae, such as this *Hydropsyche scalaris,* produce a netlike structure that filters organic material from the flowing water. This food is then grazed from the web.

Oxygen

Although relatively uniform in the atmosphere, oxygen varies markedly in aquatic habitats, primarily for five interrelated reasons:

1. Cold water has a higher saturation level than warm water (Fig. 102).
2. The oxygen level is high in the day but drops at night when plants cannot carry out photosynthesis.
3. The greater the distance from the surface, the lower the concentration when diffusion from the air is the main source of replacement.
4. Turbulent water adds oxygen at a more rapid rate than still water does.
5. The presence of high organic matter usually lowers concentrations through decay and respiration.

What effects do these characteristics of water have on insects? Species that are unable to survive lowered oxygen requirements are restricted in distribution, often to cool turbulent streams and temperate regions of the world. Adaptations for increased surface area (gills) to enhance oxygen exchange are common. Most immature insects utilize a thin cuticle that has extensive networks of tracheae and/or gills. Gills vary from fingerlike

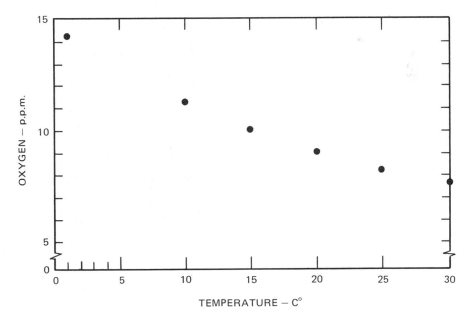

Figure 102. The maximal solubility of oxygen in water at various temperatures at sea level.

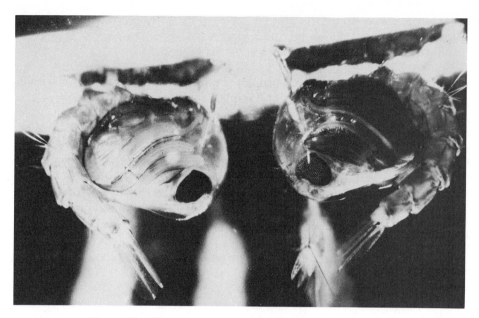

Figure 103. Mosquito pupae taking in atmospheric oxygen at surface of water through thoracic breathing tubes. Mosquito pupae, contrary to the general pattern, are not sedentary but move about in a tumbling manner by moving the terminal paddles of the abdomen.

Figure 104. Whirligig beetle (*Dineutus americanus*) with bubble serving as a plastron at posterior end of body. Although most of their adult time is spent on the surface film, these beetles are able to dive and swim beneath the surface using the bubble as an air source.

outgrowths on the abdomen (caddisfly larvae) or thorax (stonefly nymphs) to modified appendages as in mayfly nymphs. An example of internal gills, the rectal gills of dragonfly nymphs, is known.

Most aquatic adults and some immatures have open tracheal systems and come to the surface for oxygen. A common adaptation, as seen in larval and pupal mosquitoes (Fig. 103), is the location of spiracles at the end of breathing tubes from which a waxy secretion is discharged to part the surface water molecules and admit air. The water parting mechanism may be on the antenna as in adult water scavenger beetles in which a funnel of air subsequently extends back to the venter of the body and elytra to replace the depleted oxygen supply. This air bubble also serves as a *plastron,* or artificial gill (Fig. 104). As long as some nitrogen remains (it is slow to dissolve and leave), oxygen will diffuse into the plastron from the water to replace some of that utilized within the insect body. Another interesting example of oxygen uptake mechanisms is in the mosquito genus *Mansonia.* These mosquito larvae have sharply pointed breathing tubes that pierce submerged plant stems and obtain oxygen from plant tissues.

Salts

Dissolved salts influence both chemical equilibria and biological activity. With each increase comes higher water density and reduction in solubility to oxygen. Of far greater influence, however, is the effect these molecules have on osmotic pressures of cells. When salt concentrations are lower outside than within the body, as normally exists in freshwater, the body is *hyperosmotic,* and water and salts diffuse into the body, diluting blood and causing serious osmoregulatory problems. The reverse is true for salty or brackish water in which cellular fluid is *hyposmotic* and water moves from the body. Both situations are contrasted in Figure 105.

Several adaptations are present to handle problems of salt concentration. An insect that has hyperosmotic blood excretes water and ammonia in order to maintain optimal osmotic pressures. Hyposmotic individuals are similar to terrestrial forms and have cuticles that restrict salt absorption and water loss and utilize uric acid to decrease water lost through excretion.

Lakes, Ponds, and Streams

Streams and rivers have the following features:

1. currents that are directionally one-way
2. differences in flow rate that are determined by topographic slope
3. chemical variations that are usually the result of the substrate over which the water traverses
4. physical peculiarities that occur in width, depth, and bottom features.

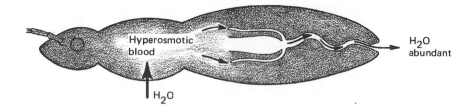

(A) Freshwater environment

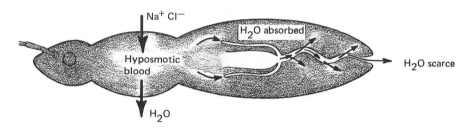

(B) Saltwater environment

Figure 105. Comparison of osmoregulation strategies. (A) freshwater environment where body water is in excess; (B) saltwater environment where water must be conserved.

Although oxygen is relatively uniform when minimal stagnation and freezing take place, this habitat becomes hazardous during spring floods. Abrasive action of suspended silt results in high insect mortality and many of the survivors are washed downstream to less favorable surroundings. Consequently, species living in streams commonly have adults that fly upstream prior to oviposition. These upstream migrations tend to nullify drift by the aquatic immatures. Within a distance of several hundred meters, the bottom, water velocity, and many other physical differences may be found, each having specific effects on oviposition and insect distribution. Figure 106 illustrates such a situation in a small stream in Kansas as seen prior to spring flooding when some of the habitat preferences become obscured. Decided differences are observable, both in species number and where maximal abundance occurs. Some insects, such as blackfly larvae and stonefly nymphs, are normally found in rapids except after flood translocation. Others, such as giant water bugs (Fig. 107) and dragonfly nymphs (Fig. 108) prefer slow-moving water. Although some species such as water scavenger beetles leave and overwinter in the soil, most aquatic insects remain active during cold temperate winters. Since a decrease of 1 °C lowers insect metabolism approximately 10 times, the same amount of food will sustain an increasing population at cool temperatures. Therefore, supplied by ample food from autumn leaves, herbivores and scavengers such as mayflies and caddisflies accomplish major growth during the winter when cold makes many fish predators sluggish or inactive. Some aquatic immatures require several years to complete development (Fig. 109).

Zone	No. orders	No. families	No. families abundant	No. restricted to zone
1	8	15	1	5
2	8	11	6	4
3	3	3	3	0
4	6	11	2	2

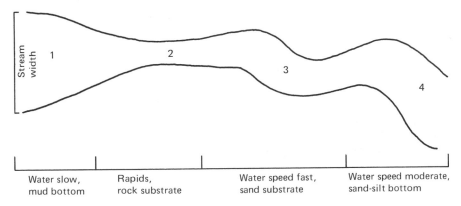

Figure 106. Characteristics of a stream in Kansas between February and April as recorded over a 10-year period. Distance represents about 200 meters. Note differences in insect taxa, their abundance, and that distribution was often restrictive.

Figure 107. Male giant water bug (*Belostoma* sp.) with eggs attached to wings.

Figure 108. This dragonfly nymph (*Anax* sp.) captures prey by means of its greatly enlarged labium. During most periods it is held in repose under the head; however, it can be quickly extended forward to capture its food as they come within striking distance. Once captured, the prey is drawn back to the other mouthparts for chewing and ingestion.

Figure 109. Dobsonfly (*Corydalus cornutus*) pupa in earthen cell under rock near stream bank. Larva migrated from the aquatic environment prior to metamorphosis.

Ponds are small bodies of water characterized by:

1. shallow depth
2. great variation in temperatures
3. early freezing and thawing
4. much diversity in dissolved gases
5. common stagnation.

Insects play vital roles in ponds, and large populations of many diverse taxonomic groups may be seen there. Most pond insects pass, or are able to pass, significant portions of their life cycle either out of water or in the basin if the pond dries up.

Since lakes are relatively large in volume, they have unique characteristics. As in streams, substrate determines much of the water chemistry, but the water depth tends to restrict both plant and insect population distribution, i.e., few insect species, mainly Diptera, are able to survive deep water and, hence, most are found in the shallow areas of the shoreline.

TERRESTRIAL ENVIRONMENT

Water

Water exists in the terrestrial region of the earth in an equilibrium between solid, liquid, and gaseous states. Below 0 °C, water is a solid, but above 100 °C temperature it exists as a gas. Evaporation can take place directly from either a solid or liquid, the rate being proportional to the temperature and concentration of water already in the air. The amount of water present when compared with the carrying capacity of air at a given temperature is termed *relative humidity*. High temperatures permit more water to remain as a gas than do low temperatures; therefore, air that has a relative humidity of 70 percent at 40 °C has more water than that that has the same relative humidity at 20 °C. When temperatures drop, relative humidity increases; if the saturation or dew point is reached, water condenses to form dew, rain, or snow, depending on the temperature and other existing environmental factors. Precipitation is not equally distributed in amount or time over the earth but varies with directional winds, their interaction, amount of moisture in the air, and presence of topographical features such as mountains. This variability, coupled with soil types, results in distinct habitats that influence distribution of both plants and animals.

Because of the evaporation properties of air, water loss from insects is normal and takes place through four major routes: (1) transpiration from body surfaces, (2) loss through breathing, (3) excretory processes, and (4) defecation. Each has been discussed previously, but a summary here may be helpful. Transpiration from the external body surface is greatly reduced by the epicuticle. Spiracular valves reduce water loss through the respiratory system; nevertheless, this accounts for approximately 60 percent of the total loss. Excretion of nitrogenous wastes requires water, but this fluid passes into the proctodeum where it is reabsorbed. Immediate restoration of lost water comes from the hemolymph where concentrations are maintained primarily by ingestion of free water, breakdown of food to release cellular fluids, and oxidation of food to produce metabolic water.

Behavior is vital in water conservation. High temperatures and wind velocities must be avoided whenever possible by seeking refuge and restricting activity. If drought conditions are normal in the ecosystem, prolonged inactivity or *estivation* often preserves body water.

In addition to having an impact on the individual, water has a vital impact on the reproduction potential of a given species. Evolution has been toward direct copulation where sperm may be transferred directly via liquids without contacting dry air. A more detailed discussion may be found in Chapter 3.

Temperature

Solar radiation, the primary source of earthly energy, passes through the atmosphere with from 80 percent to 90 percent of it being absorbed by the earth's surface. Much of this energy is radiated as heat, but transfer is slowed by the insulation of air, particularly carbon dioxide and ozone. Daytime temperatures are high because more energy is received than is radiated, but the opposite is true at night. Many animals, including most insects, depend on this radiation to maintain body temperature for normal metabolism (Fig. 110). Some scientists refer to such animals as *poikilotherms,* meaning cold-blooded, but other scientists refer to them as *ectotherms,* meaning that they obtain heat from outside the body. Ectotherm is most descriptive because the body temperatures of insects are often from 10° to 20° higher than ambient. Insects raise their internal temperature by actively exposing themselves to direct sunlight, by activity, or by seeking warm sites; conversely, they may lower their body temperature and also water loss by resting in shade and cool substrates (Figs. 111, 112). Heat gain or loss is rapid because of their small size. In most species found in the tropics and subtropics where extreme cold is rare, sunning behavior and activity can produce the necessary warming. In temperate and polar regions where great daily variation and freezing temperatures occur normally, often for extended periods, many specialized adaptations become necessary for survival. Homeostasis, the equilibrating ability, in most insects maintains temperatures higher than ambient, such as in mammals and birds, but would be detrimental in insects, for individuals the size of a mosquito would need to expend approximately 90 percent or more of

Figure 110. Short-horned grasshopper (*Melanoplus differentialis*) that has moved up the vegetation into a light wind to escape some of the summer heat.

Figure 111. American cock-roach (*Periplaneta americana*) unable to complete molt because of dry air and failure of ecdysial suture to fracture.

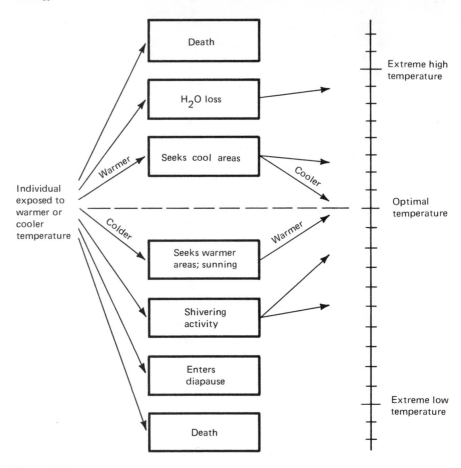

Figure 112. Some of the major homeostatic or feedback mechanisms in response to environmental temperature. Behavior is the major means of maintaining the body near the optimal temperature for insects.

their metabolic energy solely for heat. Nevertheless, there are some *endothermic* (raising body temperatures by basal metabolism) examples among insects, at least during limited periods of activity. Sphynx moths require body temperature near 36°C for flight and achieve this, during cool weather, by vibrating their thoracic muscles until sufficient heat is produced for flight. Bumblebees undergo similar behavior both before flight and during visits at flowers. Scales (in moths) and plumous setae (in bees) aid in slowing down heat loss through insulation. Nevertheless, since maintaining body temperature higher than ambient is a tremendous drain on energy, the bees must visit many flowers to replenish the energy supply.

What about insects living in temperate regions where freezing occurs regularly? Insects can protect themselves by vertical or horizontal migrations to such insulated habitats as soil or wood, by restricting activity, or by entering a facultative diapause, providing the freezing is not intense or of long duration. For species that inhabit extremely cold regions, cold resistance during some stage is vital. One such adaptation is *supercooling,* or the ability to withstand colder than freezing temperatures, often as low as -40°C (Salt, 1961). How supercooling functions is somewhat obscure, although glycerol aids in cold tolerance in many insects and excess water is removed to prevent formation of damaging ice crystals in others. Acclimation is another related factor in which exposure to gradually cooler temperatures over an extended period produces tolerance to cold extremes.

Species that inhabit hot regions face different problems. High temperatures produce increased metabolic rates, and food reserves are more rapidly depleted than at moderate temperatures. More serious, however, are the tolerance limits that, when exceeded, result in death. Upper limits are normally avoided by seeking shade, by orienting bodies so that exposure to direct sunlight is minimal, and by restricting activity to moderate temperatures (Fig. 112). In deserts, most activity occurs during the cool early morning or evening hours: Here, bees commonly collect pollen and nectar as the sky lightens before sunrise, and they cease activity several hours later because temperatures rise rapidly.

Gravity

Since air is approximately 700 times less dense than water, animals that inhabit land have problems of support. Because insects are small, their exoskeletons need not be extensive for muscles to operate effectively. Most of the adaptations to gravity have been covered in previous chapters.

Specialized Habitats

Some insects have been able to avoid the rigors of the terrestrial environment by invading plants, both living and dead. Long-horned wood borers (Fig. 113), for example, live from 2 to 3 years as larvae inside trees, literally eating their way through the wood. The environment within the resulting burrows is relatively stable, and interspecific competition is reduced. The physical state of the wood is important, as is evidence by the gradation of species preferences ranging from live trees to nearly decayed logs. Also within a given log or tree, the depth of feeding and specific tissues fed upon are important; for example, some borers prefer cambium and others prefer xylem or pith. When pupation is near, the last instar larva positions itself near the wood surface before undergoing metamorphosis. After emergence, the adult chews the remaining short distance to freedom if the larva did not make an exit.

The most common borers and miners (those that bore in leaves) are members of the orders Coleoptera, Diptera, Lepidoptera, Hymenoptera, and

Figure 113.   Long-horned wood borer and burrow. This insect is one of the few insects capable of producing its own cellulase for digesting cellulose.

Isoptera. With the exception of termites and ants, borers and miners are larvae (Figs. 114, 115). Many of these larvae are legless, have reduced eyes and antennae, and possess well-sclerotized prognathic mouthparts for chewing. The borer literally chews its way forward. Wood passing through the digestive tract may be digested by enzymes produced by the insect itself but, in most instances, protozoa or bacteria digest the cellulose, and the insect utilizes the metabolic products liberated by these symbiotic organisms (Fig. 116). In some insects much of the nourishment is not derived from the wood but from digesting fungi growing in it (Batra and Batra, 1967). Carnivorous insects also may be found in burrows feeding on borers (Fig. 117).

Feeding by most herbivorous insects causes few changes in the host plant other than destructive losses of parts. In contrast, some *monophagous* (feeding on one species of plant) and *oligophagous* (feeding on several plant species) insects evoke abnormal tissue development by their feeding or through stinging the plant. These tumorlike growths or *galls* form specific shapes and sizes, one of which is seen in Figure 118. Over 2,000 species of insects, including the order Diptera, Lepidoptera, Hymenoptera, Thysanoptera, Col-

Figure 114. Leaf miner on *Phryma* sp. As the miner grows, the width of the tunnel also enlarges.

Figure 115. Galleries formed under the bark of elm by bark beetles (*Scolytus multistriatus*). The main central gallery was made by the female as she deposited her eggs. Those radiating out from the central gallery were produced by the developing larvae as they fed.

Figure 116. The presence of borers may sometimes be determined by fecal material or frass pushed from the burrow as seen on this sunflower.

Figure 117. Some of the insects found within tunnels in wood are predaceous, as is the case with this click beetle larva (*Alaus oculatus*).

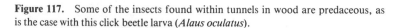

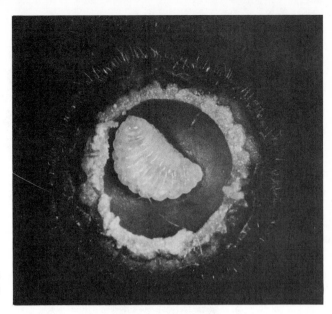

Figure 118. A gall opened up to expose the gall midge which induced its formation.

eoptera, and Homoptera (Fig. 119), in the United States induce galls. Those species, such as Homoptera, that have piercing–sucking mouthparts cannot bore out of the typical or *closed galls* and hence are found in *open galls*. Advantages gained by insects in the system include a uniform environment with much food, little water loss, and some limited predator exclusion.

Another interesting means of modifying the terrestrial environment may be seen in insects that roll or fold leaves (Fig. 120). Silk is spun across the leaf in parallel strands that shorten upon drying. The process is continued until the leaf is either folded or rolled to form a closed retreat. Some insects feed upon the leaf from within while others venture out to feed. The most common species that roll leaves are Lepidoptera. Other Lepidoptera spin silk and enclose many leaves (Fig. 121).

A further modification of leaf utilization may be seen in bagworms (Fig. 122) in which the moth caterpillar incorporates leaves into a bag of silk. This

Figure 119. Spruce galls formed by aphids (*Adelges cooleyi*) feeding.

Figure 120. The redbud leaf roller (*Fascista cercerisella*) produces a moderated microhabitat by folding over a leaf which also is used for food.

Figure 121. These fall webworms (*Hyphantria cunea*) form an extensive webbing on plants, from which they gain protection as they feed.

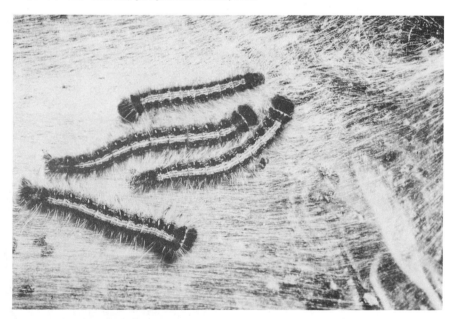

Figure 122.   A bagworm larva (*Thyridopteryx ephemeraeformis*) feeding on juniper. As the larva increases in size, the bag is also enlarged by attaching fresh leaves with silk.

enclosure can be moved about as the larva feeds. During periods of inactivity the bag is securely fastened by silk to a branch and the opening is sealed for protection. Pupation and deposition of eggs by the wingless females also take place within the silken retreat and only the winged males leave these protective retreats.

POPULATION DYNAMICS

Gene Populations

Each species represents the summation of the genes present within its interbreeding individuals, or its *gene pool*. Variation, the different expression of these genes in individuals, results from mutations in gametes and their subsequent incorporation and expression through matings and genetic recombination.

Before proceeding further, it is best to understand what occurs to a population of genes if matings are at random (*panmictic population*) and if external pressures are absent. Let us view a hypothetical population that has 30 percent of the gene pool with a dominant gene *A* and the remainder, or 70 percent, with its recessive allele *a*. If these genes segregated independently into gametes and matings occured,

(1) $p(A$ sperms$)$ $+$ $q(a$ sperms$)$ $\times$ $p(A$ ova$)$ $+$ $q(a$ ova$)$

(2) $0.30(A$ sperms$)$ $+$ $0.70(a$ sperms$)$ $\times$ $0.30(A$ ova$)$ $+$ $0.70(a$ ova$)$

then the following Mendelian genotypes and frequencies would be expected in the offspring:

(3)

Sperm

Eggs		A (0.30)	a (0.70)
	A (0.30)	AA (0.09)	Aa (0.21)
	a (0.70)	Aa (0.21)	aa (0.49)

Genotypes	Gene Frequencies
$AA = 0.09$	$A = 0.09(AA) + \frac{1}{2}0.42(Aa) = 0.30$ or 30%
$Aa = 0.42$	
$aa = 0.49$	$a = 0.49(aa) + \frac{1}{2}0.42\,(Aa) = 0.70$ or 70%

Note that under these stable conditions gene frequencies remained constant, i.e., 30 percent and 70 percent. Therefore, the genotypes and frequency, as summarized from all possible matings between genotypes, may be expressed as:

(4) $p^2AA:2pqAa:q^2aa = 1$

where $p = $ frequency of $A;$
             $q = $ frequency of $a.$

This formula, often referred to as the *Hardy-Weinberg formula,* defines a population under stable equilibrium conditions.

However, factors act upon populations that tend to place pressures on this stability. Major factors are the following:

1. *Mutations.* Changes in DNA that produce different expressions of a gene, each of which is termed an allele. Most are recessive, such as *a,* and are usually deleterious since pressure is placed on the population by adding genes that make for less fitness.
2. *Migration.* Populations are not exclusive and individuals that have

more *A* genes may enter (immigration) or leave (emigration) than those that have *a*.

3. *Nonrandom Matings.* Not every individual finds a mate.
4. *Natural Selection.* Not all individuals contribute genes equally to the next generation because some individuals are more fit to meet the environment and to reproduce more successfully than others.
5. *Genetic Drift.* In small populations, allele frequency may change because of random occurrences and may fluctuate toward the homozygous condition.

Natural Selection

Although various portions of the principle were known previously, the concept of natural selection is credited to Charles Darwin. Natural selection is based on the premise that individuals within a population have differential survival and reproductive success. There are many ways of describing the process, but the following three propositions should be sufficient:

1. Individuals within a population vary genetically. Only in parthenogenetic insect species do we find some exception to this statement.
2. Organisms have the potential to reproduce at a logarithmic rate.
3. Those individuals best adapted to the existing environment, both biotic and abiotic, survive and pass their genes on to their offspring. If, for example, the homozygous dominant condition *AA* should become *immediately* unfit when the population of *A* genes is 30 percent, then 9 percent of the potential gene population would be lost in the first generation and the gene frequency of *A* would drop to 23 percent (see formula 5).

Sperm

		A (0.30)	*a* (0.70)
	A (0.30)	*AA* (0.09)	*Aa* (0.21)
Eggs			
	a (0.70)	*Aa* (0.21)	*aa* (0.49)

(5)

Surviving Genotypes	*Surviving Gene Frequencies*

$AA = 0$ $A = \frac{1}{2}0.42(Aa) = 0.21$ or now 23%
$Aa = 0.42$
$aa = 0.49$ $a = 0.49(aa) + \frac{1}{2}0.42(Aa) = 0.70$ or now 77%

Stable populations have the best combination of genes for that given environment because of a long selection process. Under short-term circumstances, mutations and their subsequent recombinations are mostly detrimental. If the environment changes or if emigration has occured, slightly different genotypes might have increased selective advantage. The evolution of many English moths from the abundant light form, one that blended with the lichen covered bark of trees upon which they rested, to an abundance of a melanic morph or variant in response to pressures imposed by the deposition of soot upon the landscape from nearby factories is classic (Kettlewell, 1959; Bishop and Cook, 1975). This example will be discussed in the next chapter under protective coloration.

Population Density

The number of individuals of a species per unit area of space is referred to as *population density. Growth of a population* is the number of individuals added minus the number lost during a unit of time and can result from interaction of the following:

1. an increase in birth rate, or *natality*
2. a decrease in deaths, or *mortality*
3. a decrease in time to reproductive age
4. immigration into the area
5. an increase in the number of generations during the season.

The rate of growth is calculated as follows:

(6) $$\text{Growth rate } = \frac{dN}{dt} = rN$$

where r = instantaneous rate of increase;
       N = number of individuals;
       t = time.

As an increasing number of individuals initiate reproduction, the rate of increase does not remain uniform. Most populations start slowly, undergo a logarithmic increase, and then slow down as the *carrying capacity* (K) becomes critical. This can be added to produce the following formula:

(7) $$\frac{dN}{dt} = rN \cdot \left(\frac{K\text{-}N}{K}\right)$$

The resulting rate of increase per unit time produces a typical sigmoid growth curve.

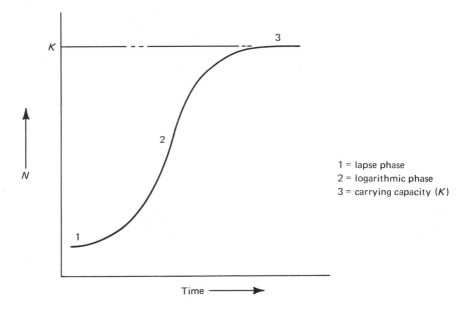

1 = lapse phase
2 = logarithmic phase
3 = carrying capacity (*K*)

When point K is reached, population growth stops and either the population crashes or a steady state is established. As this factor and others change, the size of the population will again rise or lower, thus producing periods of fluctuations in density.

What are the major factors that influence density? Previous discussions have indicated that climate has a great effect and may be referred to as *density independent,* i.e., similar changes occur whether or not the population size is 10 or 100. Many stabilizing factors, however, are *density dependent* and the number of individuals has a direct influence on population size and growth. For insects, the major density dependent factors are interrelated, but they may be categorized as *food, number of preferred habitats,* and *predators and parasites.* For example, if a pair of fruit flies is placed into a bottle with much food and optimal microclimatic conditions, reproduction will occur at a standard rate. Eventually, the population growth slows down (carrying capacity is neared) even though natality remains the same. In this instance, food has become a limiting factor at high densities where it wasn't at lower densities. Other factors may have similar effects.

Communities and Energy Flow

Individuals in insect populations do not exist alone but are part of a complex of organismic interactions. All populations of plants and animals found in a given area comprise a *community.* Such assemblages are often named for

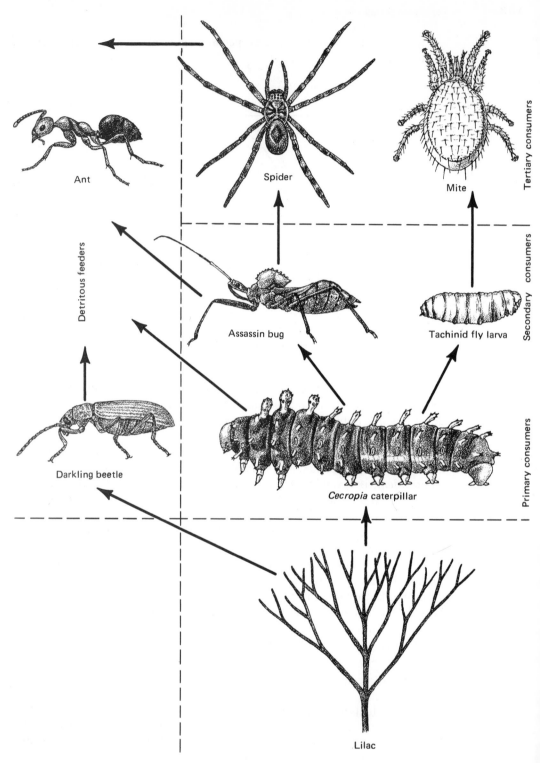

Figure 123. An example of steps in trophic levels through food chains.

the dominant plant or group of plants, such as Oak or Juniper. Boundaries between communities are often ill-defined and, hence, are arbitrary.

Plants play a dominant role in communities, as in any ecosystem, since they convert solar energy into chemical energy that enables the community to function. All other organisms are *heterotrophic,* i.e., cannot make their own food, and must feed on other organisms. Insects that ingest plants are termed *primary consumers.* These herbivores are eaten by carnivores or *secondary consumers,* which in turn serve as energy sources for other carnivores and parasites, the *tertiary consumers,* as illustrated in Figure 123. Such a sequence of trophic levels is termed a *grazing food chain* (Odom, 1971). Since most energy taken in as food is lost through cellular respiration, the amount of biomass produced at each trophic level diminishes rapidly. The efficiency of a trophic level can be calculated as follows:

(8)   Ecological Efficiency (EE) $= \dfrac{\text{calories ingested by carnivore}}{\text{calories ingested by prey}} \times 100$

Teal (1971), in his classic study of energy flow in a cold water stream ecosystem (Fig. 124), found that 2300 Kcal/m²/year of plant tissue were eaten

Figure 124. Productivity and energy flow through a coldwater stream ecosystem. Insects played a dominant role in the system energetics. All numbers represent KCal/m²/year. (Based on data from Teal, 1971.)

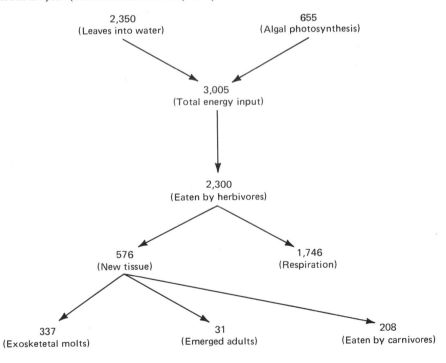

by herbivores with only 576 Kcal/m²/year becoming new tissue. Carnivores ingested 208 Kcal of this 576 Kcal/m²/year. When the 208 Kcal is divided by 2300 Kcal and multiplied by 100, an EE of 9 percent is obtained. A simplified diagram of the trophic relationships in cold water streams may be seen in Figure 125.

In contrast to the relatively simple ecosystem studied by Teal, most terrestrial communities are much more complex, and significant energy goes into producing plant support to counter gravity. Where climax vegetation consists of large trees, approximately 90 percent of the plant energy goes into synthesizing trunk and stems and little energy is available in leaves for herbivores. Eventually, however, this vast source of energy becomes available to detritus feeders (Fig. 123) and a second direction of energy flow, the *detritous food chain* (Odum, 1971).

The number of species of insects, as in all animals and plants, varies greatly with the type of community and the stage of ecological succession. Tropical rain forests have vast numbers of species, estimated at nearly 50 per-

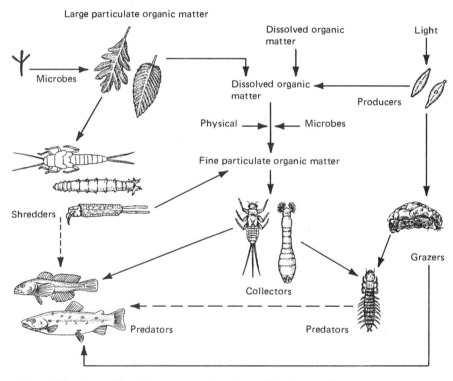

Figure 125. A simplified diagram of trophic relationships in a woodland stream. Solid lines indicate normal routes while dashes are less frequent alternatives. (Redrawn from Cummings, *Annual Review of Entomology,* 1973.)

cent of those existing, and food chains become extremely complex. Recently disturbed areas or land cultivated by humans are at the other end of the terrestrial spectrum. Here, few species are found, food chains are consequently simple, and great community instability exists.

Questions

1.   What are the major differences between the aquatic environment and terrestrial environment? How do these environments affect insects?

2.   What is a plastron? How does it function?

3.   What structures are characteristic of aquatic forms? Are these different in immatures and adults?

4.   How do aquatic inhabiting insects become distributed?

5.   If the frequency of the recessive gene t is 20 percent, what would be the frequency of the genotype TT in the population?

6.   What is carrying capacity? Express it mathematically.

7.   A pair of fruit flies is placed into a bottle with an abundance of food and optimal temperature and humidity. The population increases to approximately 200 flies but fails to go higher. What has happened?

8.   When a box of live insects being mailed from one country to another becomes damaged, the insects escape and establish a local population. Will the gene pool of this new population be identical with the original? What might happen to these two gene pools over the next 20 years? Explain your reasoning.

9.   What is an ecosystem? How does a knowledge of food chains aid one in understanding its complexity?

10.   What is ecological efficiency? How efficient are animals in converting food to tissue?

6

Behavior

In previous chapters we studied numerous adaptations that permit the organism to develop, survive, and reproduce. Directive action and reaction of an individual toward the environment using a repertoire of modifications is called *behavior.* Responses that benefit the insect permit survival and reproduction, but inappropriate behavior may result in death, inability to mate and deposit eggs properly, lowered food intake, and higher water loss. When an animal is small, the number or neurons available to coordinate potential responses is limited and much of the behavior is programmed into the system genetically. The word *instinct* has been used as a catch-all for these observed but poorly understood responses. A digger wasp illustrates behavioral sequences of this type and each response requires the previous one before being initiated, as illustrated in the diagram at the top of the next page. If the sequence is interrupted, disorientation occurs and the female wasp either enters into abnormal behavior or initiates the sequence anew.

Two types of primitive behavior are *kineses,* random movements, and *taxes,* directed movements toward or away from a stimulus. A prefix is often added to indicate the kind of stimulus being responded to, e.g., phototaxis, geotaxis, and photokinesis. Until recently, all insect behavior was considered to consist of these simple fixed-pattern responses, but, limited amounts of learning and decision making have been documented, especially in social in-

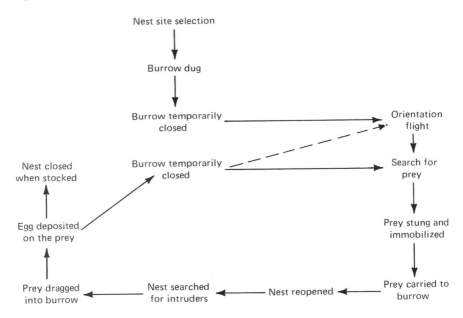

sects. Honeybees, for example, can be taught to take nectar only from certain colored dishes or from sources placed on specific patterned backgrounds (von Frisch, 1971).

A number of types of behavior are classified but only rhythms, locating food and feeding, locating mates and oviposition, orientation, and migration will be discussed because of their obvious role in survival and reproduction.

RHYTHMS

Daily or *circadian* rhythms are common in insects. The word *circadian* is derived from the Latin meaning "approximately" and "day." To indicate the precision of these responses, the term *biological clock* is used extensively in the literature. Once this "clock" has been synchronized by light (only a flash is necessary), an individual placed in constant light or dark will continue to carry out the approximate 24-hour physiological and behavioral changes characteristic of the species without further light cues. The "clock" can be altered somewhat, however, by temperature and latitude changes.

A number of behavioral patterns are closely linked to circadian rhythms. Advantages gained from such a system include such examples as initiation of molting in the early morning hours when the humidity is high, initiating activity when food sources are greatest, and synchronizing male and female development and mating periods. Insects that are active during some period of

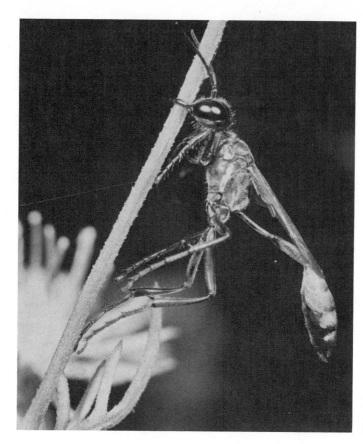

Figure 126. A solitary wasp (*Ammophila* sp.) sleeping. (Photograph by C. W. Rettenmeyer.)

the day are called *diurnal,* those at night are *nocturnal,* and those during periods of weak light such as at sunrise and sunset are *crepuscular.* Activity is often restricted to very short intervals of several hours (Fig. 126) although some may have several peaks of daily activity.

Many rhythms, in response to changes in the length of day and night, are seasonal. Some species develop during short-day periods while other insects require long-day periods. Even species that have several to many generations per year have differences in some generations in response to day length changes, often in the form of diapause (Fig. 92), reproduction, mating, or migrations.

LOCATING FOOD AND INITIATING FEEDING

Ingestion of food normally requires at least two sets of behavioral patterns. First, the food must be located. Carnivorous insects, such as mantids, are alerted by prey movements; this usually involves vision although reception of vibration may be important. Herbivores use vision or odors to locate poten-

140

tial sources. Some parasites, such as horseflies, rely greatly on vision, but others, such as fleas and mosquitos, use mainly temperature, CO_2, or host odors. Second, after the potential food has been found, a final discrimination must occur, usually utilizing sensory receptors on the mouthparts (especially the palps as in Fig. 127), antennae, or tarsi, depending on the species. Adult house flies, for example, taste with their tarsi and a positive response results in lowering of the proboscis, a second discrimination by receptors on the labellum, and finally feeding. Herbivorous insects that have chewing mouthparts normally make a trial bite to taste the food (Fig. 128), but insects that have piercing–sucking mouthparts determine suitability after preliminary probes. Not all food is equally acceptable. If the potential food is low in preference or is repelling, the insect leaves and searches anew.

Further complicating the process of feeding is the oviposition behavior of the adult female. In butterflies, for example, the proper food for the larvae is normally preselected, that is, the eggs are deposited on the plant on which the larvae will feed. Initiating feeding, however, still requires that the caterpillar taste and discriminate. If the food is nonpreferred, the individual will wander in search of the "right" food, but often starvation occurs before it is reached.

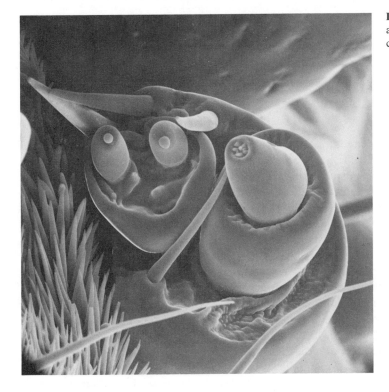

Figure 127. Maxillary lobes on a noctuid caterpillar containing chemoreceptors at their apex.

Figure 128. *Hyalophora cecropia* larva feeding on lilac. Because of its large size, the caterpillar feeds out on one side of the midrib to the tip, bending the leaf as it goes, whereupon the other side is eaten back to the leaf petiole.

LOCATING MATES AND COPULATION

The opposite sex is located by responding to at least one of a variety of stimuli including vision, hearing, smell, or touch. These stimuli normally evoke mating behavior during certain restricted time periods of a 24-hour day. Although most diurnal species use visual clues in locating mates (Fig. 129), other stimuli are necessary to initiate actual copulation. Swarming has been observed to play a major role in mayflies, caddisflies, and many primitive Diptera. These swarms, consisting primarily of males, form and orient toward conspicuous landmarks such as trees or roads. Females are attracted to and fly into these swarms where a male immediately copulates with her. Various distinctive patterns of flight, in other insects such as butterflies, initially attract mates and then color patterns and smell often are used to complete the discrimination process. Preliminary courtship may occur or mating may be carried out, however, without such behavior.

With the exception of a few species (Fig. 130), nocturnal insects require nonvisual cues for discrimination. In certain moths, the virgin female assumes a calling position and emits, from glands near the tip of the abdomen, specific

Figure 129. Compound eyes of a male horsefly (*Tabanus lineola*). The lower ommatidia are of normal size while the dorsal ommatidia are more effective in seeing objects while flying rapidly.

Figure 130. Ventral view of firefly (*Photinus pyralis*) showing abdominal luminescent organ.

sex attractant pheromones that diffuse out into the environment. These pheromones are usually specific and are detected by antennal receptors located on the male antennae (Fig. 131). If a minimal threshold of molecules (10^{-12} μg for the gypsy moth) come in contact with the antennal receptors, the male becomes "excited" and flies upwind, the most likely direction of pheromone source. Olfactory attractants bring the male close to the female where other sensory stimuli such as visual, tactile, or wing beat clues become important.

Figure 131. Pectinate antennae of giant silkworm moth (*Antheraea polyphemus*). (A) lateral view; (B) SEM micrograph of filaments showing sensory setae used in detecting female sex pheromones.

(A)

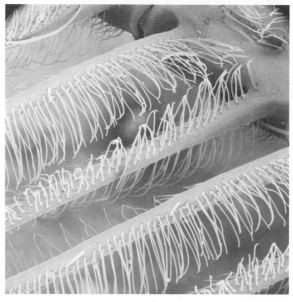

(B)

The distance of initial attraction varies. Male gypsy moths have been reported to fly nearly 4 km to a female, but effective attraction is probably less than 100 m.

In fireflies, mates are located by flashes of light from abdominal light organs (Fig. 130). Two types of strategies are involved. In the first, the female broadcasts specific signals that attract the male. In the second, the male flies about flashing its specific pattern and, when a receptive female is neared, she returns the proper flash. Each species has its own coded signal, color, shape of light organ, and height of flight to isolate species other than its own from potential matings.

Species of Orthoptera, Homoptera, and Diptera may locate mates through sound. Calls are attractive by their characteristic pitch, pulse, or intensity. In Diptera, such as mosquitoes, wing vibrations by the female produce the sound and males perceive these through antennal receptors. In contrast, it is the male in Orthoptera and Homoptera that produces the attractant sound. Long-horned grasshoppers and crickets have a *file* on the large vein of one forewing (Fig. 132) that is rubbed across the scraper at the edge of the other wing. Another means of sound production is found in cicadas where males

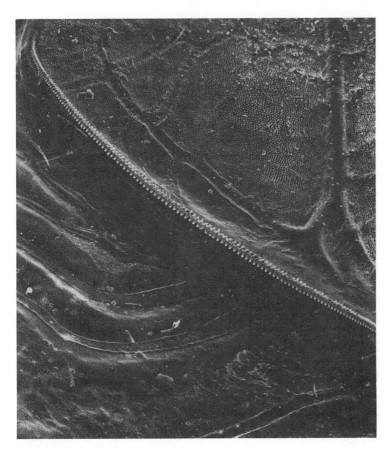

Figure 132. File on front wing of a male cricket (*Acheta assimilis*) that is rubbed across a scraper on the other forewing as the wings are moved. The result is vibration of the wings and a sound that attracts female crickets.

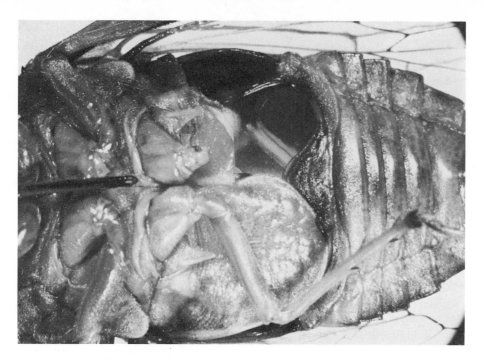

Figure 133.   Tympanic organ in a male cicada (*Tibicen pruinosa*) which produces sounds that attract the female. The right metathoracic epimeron has been removed to expose the resonant chamber of the abdomen. Each species has its particular frequency and pattern of song which maintains specific isolation.

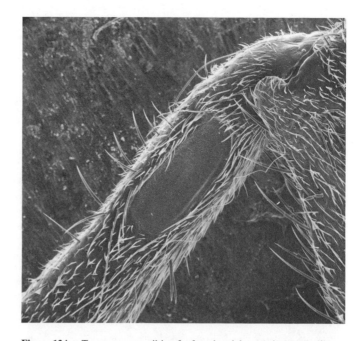

Figure 134.   Tympanum on tibia of a female cricket (*Acheta assimilis*).

possess an abdominal *tymbal* (Fig. 133) that vibrates in a resonant chamber formed from the metathoracic epimeron and first two abdominal segments. Each species has its own song and male cicadas may join together into local mating groups through synchronizing their calls. Sounds are received by the "ears" or *tympana* of the female; these tympana are located on the fore tibia (Fig. 134) in crickets and long-horned grasshoppers and elsewhere in other insects. It is interesting to note that as sound production varies with temperature, receptor thresholds also respond correspondingly, thereby insuring success of the system.

Once a potential mate has been attracted other stimuli become important to insure mating, including chemical and tactile stimuli. Certain male butterflies have *androcronia,* specialized secretory scale patches (Fig. 135), and

Figure 135.   Hind wing of male monarch butterfly (*Danaus plexippus*) with typical dark patch of secretory scales, the androcronia.

scent pencils (Fig. 136), tubular extensions from the terminal portion of the abdomen, that secrete and disperse an *attractant* or *aphrodisiac pheromone* as part of courtship behavior. In many Orthoptera, substances are produced from metanotal glands, which the female eats prior to actual mating.

Eventually, the pair attempts to mate. Many variations of copulatory positions exist (Fig. 137), especially in hemimetabolous insects. The male may mount the female or the female may mount the male. Abdomens may be twisted to fit the copulatory organs together. Once coupled, the pair may re-

Figure 136. Scent pencils on male Heliconian butterfly from Ecuador. (Photograph by C. W. Rettenmeyer.)

Figure 137. Brushfooted butterflies mating in typical lepidopteran end-to-end pattern, each facing the opposite direction.

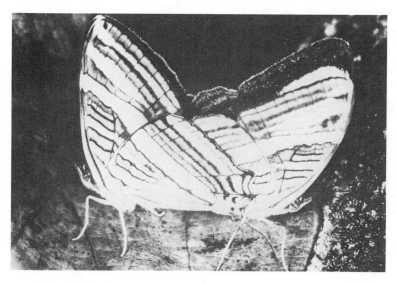

main together for less than a minute to several days, depending on the species and environmental interference. Silkworm moths have been observed to remain *in copula* for over 3 days (Fig. 200). After separating, males often search for additional mates. Females, however, have limited mating and quickly become "unattractive" to other males.

OVIPOSITION

Each species is usually very discriminating as to where eggs are deposited. Many herbivorous insects oviposit directly on the food plant necessary for development of the young (Figs. 138, 140). Similarly, parasitic species deposit eggs either on or in the host or in areas frequented by the host (Fig. 139). Some insects seem to be indiscriminate and place the ova only in a type of habitat, e.g., water or soil, but undoubtedly there is more selectivity than that observed by humans.

In most instances, at least two responses are necessary. The first involves only general discrimination to an area, shape of plant, or animal. The second

Figure 138. These *Hyalophora cecropia* eggs have been glued to lilac leaves, one of the preferred host plants for developing larvae.

Figure 139. Ichneumonid wasp (*Megar-hyssa macrurus*) with slender ovipositor to deposit egg on a burrowing insect larva. The hatched ichneumonid larva will feed on the borer host. (Photograph by R. G. Weber.)

Figure 140. Female long-horned grasshopper (*Scudderia texensis*) ovipositing in euonymus leaf. The curved ovipositor splits the leaf where feeding has occurred.

150

requires specific sensory conditions in order to initiate the behavior needed for egg deposition. Chemoreceptors and tactile receptors on the tarsi and ovipositor, tasting the substrate by the mouthparts, and other senses are employed. The long-horned grasshopper, illustrated in Figure 140, after locating the plant, made a taste bite, apparently found the plant to be suitable, and then split the leaf with its ovipositor. As the ovipositor was withdrawn, the egg was released. During the entire oviposition, the female continued to feed. Face flies locate manure by smell, alight, and sense the feces through receptors on the feet, labellum, and ovipositor (Fig. 54). If the feces have the proper moisture content, physical constitution, and aroma, eggs are laid.

Some of the more complete studies on oviposition have dealt with solitary wasps whose fixed action patterns are often inflexible and must be performed sequentially. The female usually digs a burrow in the ground, temporarily closes it, and then searches for some specific prey. Those that seek caterpillars sting them in every body segment to prevent segmental reflexes and bite the head of the prey to prevent jaw movement. Others subdue crickets or spiders in very precise patterns. The prey is now carried or dragged to the burrow entrance, the nest is opened and searched, the prey is pulled inside, and an egg is deposited on the prey. The sequence is repeated until the nest is sufficiently stocked for future larva to complete development. Each of the previous sequences requires specific stimuli to initiate, and an interruption often results in either repeating the behavior or starting the entire sequence over again. For example, the artificial introduction of prey at the wrong time will usually evoke a behavior necessary to repel invaders instead of initiating a predatory pattern.

ORIENTATION

Little is known about orientation, but major landmarks such as the sun and moon serve as important reference points for many insects. Most of the limited research has been done with Hymenoptera. Many wasps take short orientation flights around the nest before flying off in search of food. Objects such as trees, highways, rivers, and large rocks are used to fix their position. In addition to these landmarks, the sun is also used as a reference point by many bees and ants. The insect compensates for the movement of the sun across the sky and is able to return to the nest by using this reference guide. The direction of food sources may be passed on to other honeybees in the hive by means of a dance in which the sun is used as the reference point (Fig. 189).

Moths use the moon to orient themselves. Normally, a certain angle between the eye and the distant moon is maintained during flight. If a moth approaches a light or fire, however, it attempts to maintain this angle to the new bright "moon" and starts a circling motion about the light (Wigglesworth, 1972).

MIGRATION

Unlike birds, relatively few insect species engage in dramatic geographical mass movements each year. Those that do are usually characterized by:

1. undergoing this behavior soon after adulthood is reached
2. females with developing ovaries being the consistent migrant sex
3. the migration often being a one-way movement although some of the best known examples, such as certain populations of the monarch butterfly (Fig. 141), do make at least partial return flights.

Migrations often start in an area where populations are increasing and where environmental carrying capacity pressures are building. Under these conditions, nymphal *Schistocerca gregaria,* a Middle East grasshopper, becomes morphologically and behaviorally differentiated into a migratory phase. When these grasshoppers reach adulthood, there is an exodus flight that normally encounters and gains assistance from winds. Big swarms may extend over 250 sq km. Flight of these and other migratory insects is maintained

Figure 141.   Monarch butterflies (*Danaus plexippus*) often migrate great distances. This specimen has been tagged as to its release point to determine distance, duration, and direction of flight.

until other stimuli, e.g., matured ovaries, wing muscle deterioration, energy exhaustion, length of day, and other settling responses redirect behavior.

A major function of migration is dispersal of individuals to new and suitable habitats. In many cases, since these new areas are not able to support a species population during the entire year or the climatic conditions won't permit overwintering, reinvasion occurs each year.

Johnson (1966), using the adult life span as his major criterion, lists three types of migrations: *Type I* is for "short-lived adults that emigrate and die within a season" (some locusts, butterflies, and aphids). *Type II* are "short-lived adults that emigrate and return" (dragonflies). *Type III* are long-lived adults that migrate before or after hibernating or aestivating and may or may not return (some beetles and Noctuid moths).

CONCEALING COLORATION

Most insects are difficult to see. They blend with their background and, unless movement (Fig. 142) betrays their position, go unnoticed. This is called *concealing* or *cryptic coloration*. A number of grasshoppers are green and are

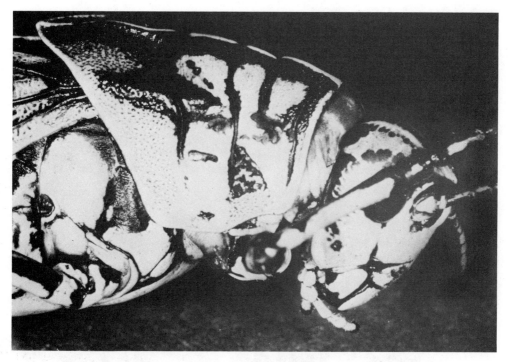

Figure 142.   Preening behavior of short-horned grasshopper (*Romalea microptera*) using its tibia to rub off dust particles from its compound eye.

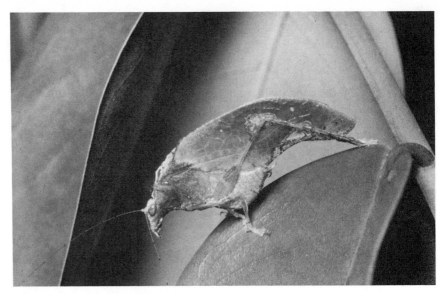

Figure 143. This long-horned grasshopper (*Pycnopalpa bicordata*) is somewhat conspicuous here but when located on leaves that possess fungal infections this insect is difficult to observe. (Photograph by C.W. Rettenmeyer.)

Figure 144. A walking stick effectively hidden. (Photograph by C. W. Rettenmeyer.)

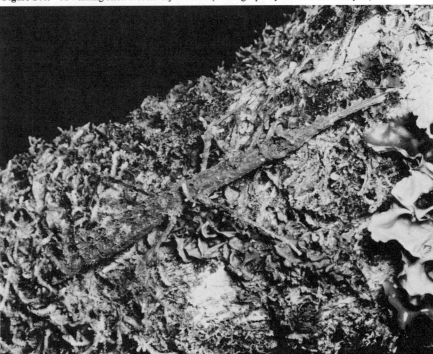

Figure 145. Butterflies (Satyrid and Ithomiid) from Costa Rica that are transparent and, with their slow flight next to the ground, are very difficult to see.

difficult to distinguish from leaves. Exceptional examples may be seen in the long-horned grasshoppers in which a few species even have brown areas on their wings that resemble disease patterns commonly found on host leaves (Fig. 143). Others, including walking sticks and most mantids, resemble twigs (Fig. 144). Some treehoppers resemble thorns. A few butterflies in the tropics have transparent wings and continuously blend with the background while resting or during flight (Fig. 145). Certain walking sticks change color from day to night thereby blending more closely with their background. Countershading (one side darker than the other) may be present and, from the side, results in a more flattened appearing object.

Inherent in the concept of concealing coloration is the axiom that cryptic coloration is protective only when the insect rests upon the proper background (Figs. 146, 147, 148, 149). Experimental data during the past ten years suggest that insects do, in fact, search for suitable substrates. Moths that have stripes tend to orient themselves to correspond to lines on bark upon which they rest. Both tactile and visual stimuli are involved for this orientation.

Does cryptic coloration actually provide protection from predators or is this anthropomorphic reasoning? Evidence exists that it does provide protection, at least from vertebrate predators. Birds often pass by insects that are shaped and colored like twigs, leaves and feces (Fig. 148), until one of the insects moves, whereupon the bird seems to become programmed and locates ad-

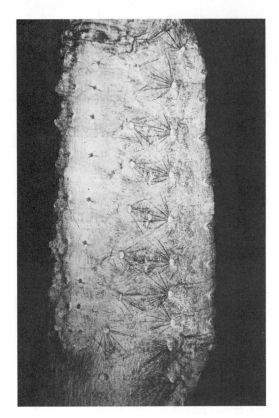

Figure 146. Protective coloration in the noctuid caterpillar (*Catocala* sp.) whereby the larva blends with the background by color and the form disruption produced by the long setae at the body sides.

Figure 147. Butterflies are not as commonly cryptic on the dorsal surface as are moths. This brushfooted butterfly, however, blends in with trees with light bark, upon which it rests with its head in the downward position.

156

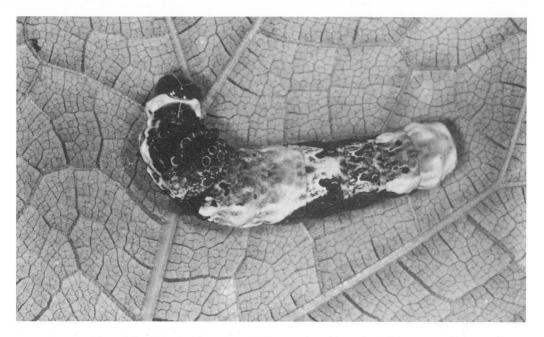

Figure 148. Some lepidopteran caterpillars, such as this swallowtail larva, resemble bird droppings and predators apparently pass by them without recognizing the larvae as food.

Figure 149. Dark and light forms of the peppered moth (*Biston betularia*) were photographed on the trunk of an oak blackened by the smoke-polluted air of the English city Liverpool. The light form is clearly visible whereas the dark form is almost invisible below and to the left. (Photograph by J. A. Bishop, University of Liverpool.)

ditional insects with ease. As a result, some birds may attack any object resembling this gestalt form, including twigs and leaves. If similar insects become scarce or if too many vegetative objects are encountered, the bird usually reverts to its original attitude and again passes by camouflaged insects. Adult giant silkworm moths, and probably most large insects with cryptic coloration, also tend to have short adult survival periods, apparently a selection that lowers exposure of the total population to predation (Blest, 1963; Ricklefs, 1973).

As in all animals, differences exist in color and patterns within the same species (polymorphism). Very often variability includes distinct dark and light individuals, which brings up the question of what are the mechanisms involved. What possible advantage can come from not blending with the background? Certainly the benefits would not seem to be to the individual easily located by a predator. Observations on the English peppered moth, *Biston betularia* (Figs. 149, 150) indicate that both light and melanic forms attempt to select suitable background color. Before industrialization, light forms were the most successful because of the numerous trees covered with lichens while the dark ones had fewer suitable substrates to rest on and, hence, were more easily seen and eaten. With the increase of industrialization and subsequent deposition of soot on the landscape, lichens became sparse or disap-

Figure 150. Same two forms as in Figure 149 of the peppered moths were photographed where less soot and more lichens are present. Here it is the dark form which may be clearly seen. The light form is below the dark form. (Photograph by J.A. Bishop, University of Liverpool.)

Figure 151.   Larval leaf beetles *(Lema trillineata)* with fecal matter attached to aid in concealment.

peared and the light phase then became the conspicuous form. *Industrial melanism,* as the previous process is termed, resulted in a reversal of selection advantage, and, in addition to *B. betularis,* approximately 70 of 780 other English moth species evolved toward predominantly darker phases. A return to the original state is possible, however, and has occurred recently in a few areas of England where laws have drastically reduced industrial pollution and where the landscape is returning to its original state (Bishop and Cook, 1975).

In addition to morphological adaptations, some insects conceal themselves with available material from the environment. Certain leaf beetles (Fig. 151) cover their bodies with fecal material during their larval stages and become less conspicuous. Caddisflies, also as larvae, use pebbles and twigs from streams to form cases that blend with the background and also provide physical protection (Fig. 280). A caterpillar in Borneo attaches flower buds from its food plant to its body and replaces the buds as they become wilted.

REVEALING COLORATION

Some of the larger insects have forewings cryptically colored but the hindwings are brightly colored or contain eyespots. When resting, the forewings cover the body and hindwings and provide concealment. Such cryptic coloration is of no benefit once discovered, and birds, upon locating such prey, initially peck and often seize the insect by the head for easy ingestion. At

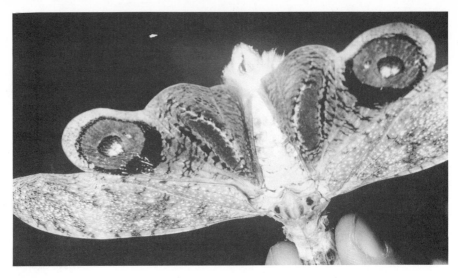

Figure 152. Many of the large tropical insects, such as this peanut bug (*Lanternaria phosphorea*), have either bright colors or eye spots on their hind wings. When grabbed by the head, as commonly occurs when birds attack, these wings are exposed and often startle the predator sufficiently to permit escape.

this point, the bright hindwings or eyespots are exposed by reflex action (Fig. 152) and startle the bird, often causing the insect to be released. Examples of this type of wing coloration may be seen in many giant silkworm moths, hawkmoths, some grasshoppers, and a few tropical mantids and planthoppers.

In some butterflies, the dorsal surface of the wings is brightly colored and iridescent, whereas the ventral surface is drab (Fig. 153). Light may be

Figure 153. The genus *Morpho,* a tropical group, includes many species that are iridescent on the dorsal surface whereas the ventral surface is cryptic or drab. The flash effect as this butterfly flies makes capture difficult.

reflected great distances from such iridescent wings, often observable from low flying airplanes. During the flight of these butterflies, the alternative bright–dull–bright sequential appearance of the wings is deceptive and few predators are successful in their attempts to capture this prey.

How does this coloration aid in survival in uniformly bright colored insects (Fig. 154), especially since birds have the ability to see red? Evidence has been accumulating that brightly colored insects are often rejected by vertebrate predators. If such species were unpalatable, a "linkage" of this trait with color would be mutually beneficial to bird and potential prey. The color would serve as a warning and the predator would not waste time and energy on nonpalatable forms. Until recently, little experimental evidence existed to verify that some brightly colored insects are, indeed, distasteful and the coloration *aposematic* (warning) to vertebrate predators (Sexton, 1960; Brower et al., 1963; Holling, 1965; Brower, 1969). Distastefulness may be the result of the production of internal substances, as in the majority of insects, or from incorporation of ingested toxic materials. Milkweed plants often contain cardiac glycosides (similar to digitalis) which cause vertebrates to vomit and eliminate

Figure 154. Aposomatic notodontid larvae (*Datana ministra*) often assume this characteristic pose when disturbed.

Figure 155. This monarch larva (*Danaus plexippus*) feeds on milkweed and incorporates toxic secondary plant substances, when present, into its body. Protection from predators is thereby gained using the host plant's biochemistry.

toxins from their bodies prior to serious damage. In contrast to the vertebrates, certain butterfly larvae, including the common monarch, use these plants exlusively for food. Experiments by Brower (1969) indicate that monarch caterpillars (Fig. 155) incorporate these secondary plant substances into their body and retain them into adulthood. The result is a butterfly that has conspicuous coloration and is unpalatable. Brower found a strain of monarch that could be fed cabbage and, therefore, lacked the toxin from milkweed. Blue jays were trained to feed upon these palatable butterflies; then monarchs reared on a *Asciepias curassavica,* a milkweed containing toxins, were introduced. After eating those reared on milkweed, the birds became violently ill, vomited many times over a half-hour period (Fig. 156), but all

Figure 156. A jay vomiting after ingesting nonpalatable monarch butterfly. (Photograph by Lincoln B. Brower.)

Figure 157. Swallowtail larvae have an osmeterium which is everted when disturbed and releases disagreeable odors. This larva is *Papilio polyxenes.*

recovered. Subsequently, these birds refused monarchs on sight. Although not all North American milkweeds are toxic, apparently only a limited number of butterflies produced from toxic plants must be available (30 percent or more) for selective advantage to occur. The higher the density of unpalatable individuals, the greater the protection.

Distastefulness is not the only factor for conditioning vertebrate predators. Wasps and bees are commonly avoided because of their stings, not because of their unpalatability; if the stings are removed, unconditioned birds readily eat these insects, but the birds will avoid these insects once they have been stung. The birds associate the sting with the color pattern. Insects with aposomatic coloration are also avoided because of noxious odors, such as those produced by anal glands in certain ground beetles and osmeteria of swallowtail butterfly larvae (Fig. 157), or of stinging hair or setae, sometimes found on certain lepidopteran caterpillars (Fig. 205).

MIMICRY

Observers have noted that the number of bright color patterns is limited. Once an unpalatable taste or sting and color pattern have been established, convergent evolution may take place between other insects with similar patterns. *Mimicry* occurs when a species of animal (*mimic*) resembles another animal species (*model*) which is avoided thus gaining protection because of previous predator conditioning (Figs. 158, 159). Behavior, as well as mor-

163

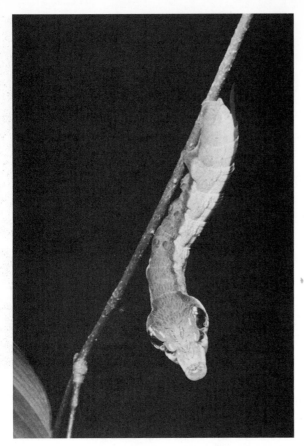

Figure 158. A sphingid larva which mimics a snake, when disturbed, by swelling the anterior region of the body and swaying back and forth. (Photograph by C. W. Rettenmeyer.)

Figure 159. A hemipteran mimic of ants. Note the reduced wings, elbowing of the antennae, and the narrowing of the abdomen for the close resemblance.

phology, is involved with the mimic flying or running like the model (Figs. 160, 161). The mimic is usually less abundant and exists in the same habitat as the model. Whenever the mimic becomes more abundant or the model less numerous, then selection pressure is increased on both groups.

Although mimicry has been categorized into many types, only the two most common will be discussed. The first type is termed *Batesian mimicry* after the noted English naturalist, Henry W. Bates. Here, the mimic is restricted to those species that are palatable but gain benefit because predators are apparently deceived. The classic example of Batesian mimicry is the

Figure 160. A long-horned grasshopper (*Aganacris insectivora*) which is a mimic of a spider wasp. The antennal tips are black to give the appearance of shortened length. (Photograph by C. W. Rettenmeyer.)

Figure 161. A reduviid or assassin bug which is a mimic, as in Figure 160, of a spider wasp. (Photograph by C. W. Rettenmeyer.)

Figure 162. The lower two butterflies are the models, the monarch, while the upper is the mimic, the viceroy (*Limenitis archippus*). Since the viceroy is palatable, the relationship is termed Batesian mimicry.

monarch (model) and the viceroy (mimic) butterflies. Both species have similar color patterns (Fig. 162), and birds, having learned to avoid the monarch, as seen previously, also avoid the viceroy. An untrained bird may feed on numerous viceroys and palatable monarchs (those that have developed on milkweed without toxins) until encountering its first unpalatable model, after which further predation upon this color pattern ceases (Brower, 1969). Sexton (1960) studied the firefly *Photinus* and recorded the correlation between the pronotal and elytral colors and distastefulness. Brower et al. (1963) concluded that phylogeny of certain helioconian butterfly species was correlated with palatability to silverbeak tanagers. Hundreds of other examples of apparent Batesian mimics may be found in the literature, but few have been scientifically substantiated.

The second type of mimicry is termed *Mullerian mimicry,* named after Fritz Muller. Here both model and mimic are unpalatable and the ingestion of either by the vertebrate predators results in the avoidance of both species. Whereas Batesian mimicry is usually disadvantageous to the model, Mullerian mimicry results in benefits to both mimic and model since both are unpalatable.

Brower et al. (1963) studied relative palatability of helioconian butterflies to silverbeak tanagers in Trinidad and noted that similar shape and size gave selective advantage to the Mullerian complex of butterflies. Holling (1965) indicated that the higher the density of models and mimics in a Mullerian complex (Fig. 163), the greater reinforcement on the predator search image and the greater the survival benefits until other density-dependent factors become dominant. Holling also noted that total rejection of the distasteful prey is not necessary to raise the predator attack behavior threshold and that, contrary to common belief, the mimic can outnumber the model in this type of mimicry. In addition to butterflies, common examples of Mullerian mimicry can be seen among the lycid beetles, wasps, bees, and ants.

Figure 163. Examples of similar color patterns and mimicry in six species of butterflies in Costa Rica.

Questions

1.   What is instinctive behavior? What benefits are gained by instinctive behavior over requiring learning? Does the size of the organism have any effect upon its ability to learn?

2.   What advantages are gained by circadian rhythms?

3.   Are there behavioral differences between locating food and initiation of feeding?

4.   Does behavior prevent interspecific matings? Explain your answer.

5.   What are some of the advantages and disadvantages of utilizing insects as experimental animals in elucidating animal behavior?

6.   What are the advantages and disadvantages to concealing or cryptic coloration and revealing coloration? Observe some insects and attempt to determine why they are colored as they are.

7.   Are there advantages for two species to have similar aposematic coloration? Explain your conclusions.

7

From Solitary to Social

All animals are characterized by

1. irritability, or the ability to sense and respond to stimuli
2. locomotion
3. metabolism, the chemical reactions within protoplasm
4. growth, an increase in mass as a result of metabolism
5. adaptation, or the ability to survive in a changing environment
6. reproduction

These fundamental life processes exist in single-celled organisms and are retained by the complex multicellular organisms.

Most insects exist as *solitary* animals, i.e., they live as separate individual entities (not social). Each organism is primarily concerned with satisfying the biological requirements for self-survival. Interaction with others of the same species usually occurs by chance and is not normally "intentional" except for certain periods of mating. Parents normally leave soon after egg deposition (Fig. 164); therefore, each individual is independent from the very beginning (Fig. 165) and its only parental legacy may be the specific environment into which it was deposited and a nearby food source.

In some species of insects, however, individuals associate with one another during certain stages that cannot be explained by chance meetings as a

Figure 164. Rearing cells constructed of mud by the solitary organ pipe wasp (*Trypoxylon* sp.) with exit holes by wasps and their parasites. Although many cells, each to feed a developing larva, are mass provisioned by a female, there is no gregariousness exhibited by emerging wasps.

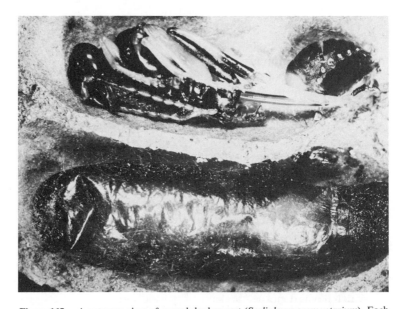

Figure 165. A cutaway view of a mud dauber nest (*Sceliphron caementarium*). Each cell contains a pupa. The pupa at the top has had the last larval exoskeleton removed.

result of high populations, eggs being deposited together, or ecological conditions causing localization of individuals. Certain caterpillars have been shown to have an aggregation behavior and are gregarious during this immature stage (Fig. 166); they become typically solitary insects as adults. Other forms of gregariousness may be seen in certain wasp sleeping aggregations, hibernating aggregations, and migratory aggregations of certain Lepidoptera. Many of the latter two are probably the result of limited microhabitat and macrohabitat suitability and are not true gregarious behavior.

In some rare instances a species may stay and care for the eggs or young for a short period of time, as is the case of many earwigs. Passalid beetles have all stages living together in rotting logs, and certain dung beetles, after both male and females prepare a nest, remain with the immatures until the larval development is complete. Some solitary bees may share a common entrance hole but build separate nests. In the previous examples we see circumstances

Figure 166. Gregarious caterpillars in a Costa Rican rain forest.

that involve prolonged interactions between individuals, a prerequisite necessary for socialization.

Insect societies have arisen independently several times within the orders Hymenoptera and Isoptera. What factors enhanced such evolution? Michener (1969) proposed that eusocial bees evolved from either *subsocial* groups, family groups in which the female rears the young but dies or leaves before maturity in the young, or from a *semisocial* group, a family of females of the same generation who cooperate but are not necessarily related. Hamilton (1964) suggested that societies were the result of kin-selection on the haplodiploidy form of reproduction (males = haploid, females = diploid) and that there is a selective advantage for daughters to help their mother rear what would be their sisters. Daughter siblings are more genetically related to one another under haplodiploidy than they are to their mother (Hamilton, 1964; Wilson, 1971; Eberhard, 1975). Hamilton's concept broke with tradition in that selection for societies was also on the group level by altruism as well as on the individual. Alexander (1974) proposed that kin-selection benefited the parents rather than the siblings. Eberhard (1975) reviewed the varying theories of social behavior and suggested that each wasn't exclusive but that Michener's mutualism, Hamilton's kin-selection, and Alexander's parental manipulation were interrelated or functioned sequentially in colony and sterile worker evolution.

The formation of colonies "frees" most individuals from expending energy for reproduction and permits a division of labor or specialization that greatly benefits the entire society. Attempts have been made to compare the individual from such a colony with the cell and the colony itself with the multicellular body of an organism; this "superorganism" concept is intriguing, but it has fallen by the way and is not looked upon with favor by most biologists.

Much can be learned by contrasting primitive societies with highly evolved societies. When such comparisons are made, the following evolutionary trends appear:

1. production of casts of individuals that are behaviorly (polyethism) and usually morphologically (polymorphism) specialized for various functions within the colony, e.g., workers, soldiers, reproductives
2. increased population size within the colony
3. daily provisioning or direct feeding of the young (Hymenoptera)
4. production of a nest or physical structure to enclose the colony that can moderate environmental conditions
5. complex variations of trophallaxis or mutual exchange of substances from the proctodeum and/or salivary glands
6. colony odor whereby individuals become restricted in their activity and normally cannot enter and join other colonies

Figure 167. Thysanuran (*Trichatelura manni*) in an army ant column of *Eciton burchelli*. Because of the nomadic behavior of the ants, which move daily, the thysanuran must be and is capable of following the trail of the ant migration. (Photograph by C. W. Rettenmeyer.)

The success of insect societies is apparent. The major predators and scavengers in many areas are social insects.

Along with highly evolved insect societies come opportunistic organisms; many thousands of arthropod species have become associated with social insects, especially termites and ants (Figs. 167, 168). Some of these *inquilines,* or guests, e.g., Coleoptera and mites, are tolerated and in some cases defended by the societies and through time have come to resemble the hosts in both structure and behavior. Other inquilines are scavengers or are predaceous on the host population or are essentially social parasites with the host ants that feed and rear them. Certain insects, such as a few Homoptera, are mutualistic and provide some nutritive substances for the ants; these *trophobionts* are protected by their hosts and are analogous to domestic cows, for they yield food sugar solutions from the hindgut, "honey-dew," upon request.

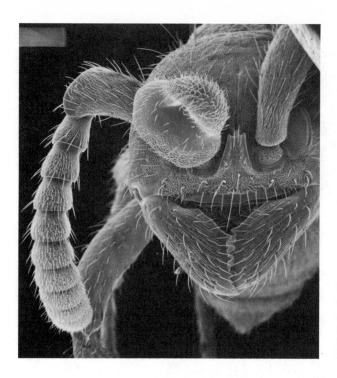

Figure 168. Mite (*Antennequesoma* sp.) attached to antenna of army ant (*Nomamyrmex esenbecki*). The "fit" is only for the antennal site, and the color and ornamentation are very similar to the ant host.

ISOPTERA (TERMITES)

There are more than 2,000 (over 50 in the U.S.) species of termites, all of which are social and form colonies from several hundred individuals to over a million in some of the highly specialized species. Termites are unique among social insects in that (1) they undergo incomplete metamorphosis with the nymphs, after the second or third instars and at least in the primitive species, serving as workers in the colony and (2) members of the colony usually consist of nearly equal numbers of both males and females, each playing an apparently equal role in the social structure.

Termites are believed to have evolved from cockroaches. Evidence for such ancestry includes observations that certain wood-boring cockroaches live in aggregations with all stages of development present, feed on decaying wood, and require intestinal protozoa to derive benefits from cellulose. Close relationships between these cockroach protozoa and those found in primitive termites strengthen this hypothesis (Krishna and Weesner, 1969). From this primitive type of association, termites have evolved into the complexity of social and morphological structure seen today with most individuals being sterile, seldom or never exposed to the outside environment in temperate regions, and restricted to specific roles or castes that insure the survival of the colony. Each colony is based on *reproductives* that probably reflect the ancestral solitary or subsocial form from which termites have arisen. Other castes include *soldiers* and *workers* (Fig. 169). *Pseudergates,* last instar worker nymphs in primitive species, retain a potential to molt into different castes (Krishna and Weesner, 1969).

New colonies result from joint effort of both a reproductive male (king) and female (queen). At appropriate times of the year, winged reproductives swarm from the nest. Following a dispersal flight, wings are lost, a male and female court, and then proceed to locate a suitable nest site. Next, a chamber or *copularium* is formed, the pair mate, and after several weeks to a month, a dozen or more eggs are deposited. Eggs are cleaned meticulously to prevent fungal infection, and after several weeks the eggs hatch. Early instar nymphs feed upon fecal material, one form of trophallaxis, which infects them with symbiotic protozoa or bacteria. In some primitive species, these nymphs serve as the only workers within the colony. In the higher termites, however, a true worker caste of sterile adult workers develops and nymphs carry out only minor activities. Other nymphs (usually both males or females) metamorphose into soldiers. All soldiers have hard heads and protect the colony either by large mandibles (Fig. 169) or by squirting sticky fluids from a specialized fron-

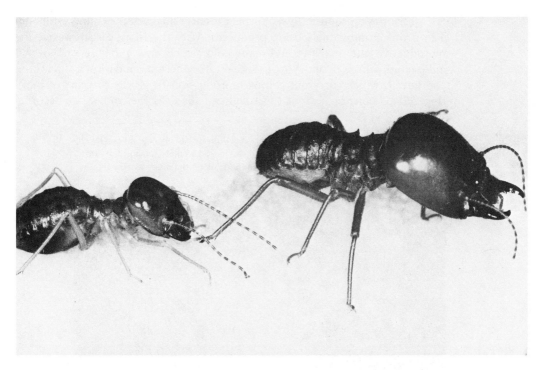

Figure 169. Two castes in a South American subterranean termite. The worker is to the left, the soldier to the right. This soldier's mandibles can easily cut through human skin and draw blood.

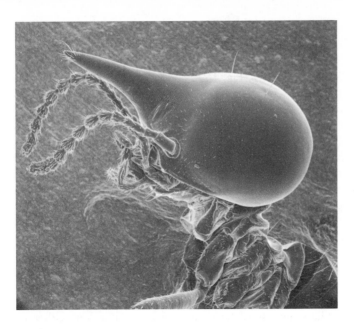

Figure 170. Scanning electron microscope micrograph of a *nasute*. This specialized soldier is able to eject sticky and noxious fluids from the frontal process and is especially effective on ants, the major predator of this species.

tal protuberance (Fig. 170). After the colony has matured (from 2 to 4 years), winged reproductives appear and normally leave the colony when climatic conditions are favorable.

Termites reproduce slowly during the early stages of the colony with only 4 or 5 dozen individuals being produced. After the first year, however, the reproductive rate increases and many colonies become immense with over 2 million individuals in one African species. Most termites have but a single king and queen, which carry out colony reproduction for 10 to 20 years or more (Fig. 171). If one or both of this royal pair die or if the colony size becomes

Figure 171. The greatly enlarged (*physogastric*) queen has become an egg-laying machine in her enclosed chamber. Note the specialized soldiers, or nasutes, which are capable of squirting a sticky secretion upon invaders, the many worker retinue, and the young nymph above the queen.

large and dispersed, reproductive replacements may come from either reproductives remaining within the colony and not swarming or from developing immatures, pseudergates, becoming fertile adults or *supplementary reproductives* (Fig. 172). These begin reproducing within 6 to 8 weeks after a colony has been deprived of the queen or king by isolation or disaster, and they enable the survival of the unit.

The mechanism for caste determination is not completely understood but current evidence indicates that it is under the influence of at least two pheromones. One type of these "social hormones" is produced by the royal pair, transferred throughout the colony by fecal trophallaxis, and apparently inhibits the production of secondary reproductives from pseudergates, as seen in the dry-wood termite, *Kalotermes flavicollis* (Fabr.) (Luscher, 1961). The other is produced by soldiers and, when a certain concentration exists in a colony, inhibits the development of most of the nymphs beyond the nymphal worker stage to soldiers. A sudden decline in soldiers because of death from an ant invasion, for example, would have an immediate effect on developing nymphs, and the proper balance of castes would be restored.

Food for workers normally consists of cellulose (such as in grass, leaf litter, humus, and paper) although some utilize fungus growing on the plant mat-

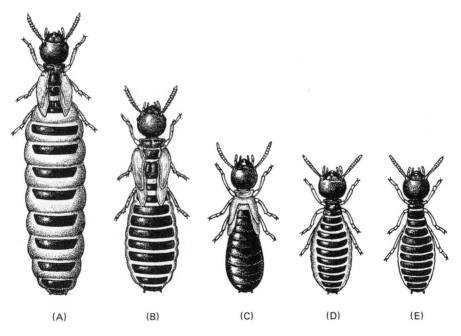

(A)   (B)   (C)   (D)   (E)

Figure 172. Certain termites with extremely large colonies, such as this African species (*Amitermes hastatus*), may have supplementary reproductives. (A) secondary queen; (B) secondary king; (C) antepenultimate nymphal instar; (D) tertiary queen; (E) tertiary king. (Redrawn from Skaife, 1954.)

Figure 173. A dissected tube from a termite nest (*Reticulitermes flavipes*) to the wood in a house in Kansas. This tube permits the termites to return to the soil, necessary in this species, and yet have access to the wood without encountering the desiccating action of the dry air.

Figure 174. Rather than using a tube, this South American termite colony uses a mud sheet to cover the food source and route to nest.

ter brought into the nest. Most soil-inhabiting termites have burrows or tunnels to the food source. If the food is wood and is not in contact with the soil, these tunnels may be extended to the source by means of tubes formed from cementing soil and wood particles with intestinal secretions and saliva (Figs. 173, 174). If the food consists of grasses, such as in some African and Australian species, foraging may be carried out on the ground surface at night or through underground tunnels. Cellulose is digested by cellulase enzymes from protozoan symbionts in primitive species and bacteria in advanced forms. Subsequent digestion of the released sugars by the termite host produces the necessary metabolic energy, but proteins and vitamins probably come from either digesting the intestinal symbionts or by cannibalism (which is relatively common). Workers then feed the soldiers, reproductives, and early instar nymphs by trophallaxis; the more common type is probably salivary exchange which often contains wood fragments, but proctodeal releases of wood fragments and symbionts are also needed. Reproductives receive salivary secretions only.

A variety of nests are constructed, varying from random shapes in soil forms to oval nests on trees (Fig. 175) in tropical areas to elaborate ground

Figure 175. A typical termite nest in South America. These *Nasutitermes* nests are usually found in trees and manufactured from wood and soil particles mixed with termite secretions.

Figure 176. Mud termite nests in South America. These termites forage for grass and, when abundant, seriously compete with vertebrate herbivores on pastures.

nests (Fig. 176) in semiarid regions, some of which in africa may extend upward 20 ft (6m) and measure 12 ft (3.66m) in diameter. These latter *termitaria* are produced by cementing soil and fecal material with secretions from either specialized glands in the termite's head (frontal gland) or from the proctodeum. Termitaria are veritable fortresses and often survive as geographic landmarks for years after the colony has died. Because of their unique construction (Fig. 177), these nests are also able to control water loss (often the

Figure 177. A cross section of one of the mud termite nests in Figure 176. Note the many tunnels, the central core where the queen and young are located, and the bottom cavities or basement.

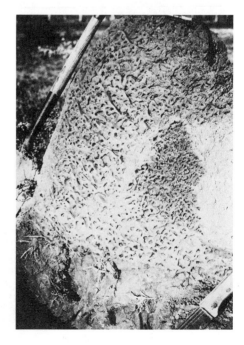

most critical restriction of termites) while still providing adequate ventilation and temperature regulation.

HYMENOPTERA (SOCIAL WASPS, ANTS, SOCIAL BEES)

The evolution of societies has occurred on at least 11 independent occasions (Wilson, 1971). These societies are:

1. predominanly female
2. normally restricted to reproductives and workers
3. haplodiploid in sex determination with males resulting from parthenogenesis and only functioning in reproducing
4. characterized by legless larvae that must be fed at least once a day by the adults

Considerable colony energy, therefore, must be expended toward rearing larvae that are not able to contribute significantly to colony labor.

Social Wasps

The least specialized social Hymenoptera are found among primitive wasps whose colonies are small, often numbering only several dozen individuals. Each colony may have several to many queens although one queen is normally aggressively dominant and the remainder function as workers. A sterile worker caste, when present, is difficult to distinguish from queens and consists of few individuals. Nests consist of an open layer of brood cells suspended by a single pedicel (Fig. 178). In the temperate region, each colony survives only for several months during the summer (Fig. 179) and disorganizes during autumn.

In the more advanced social wasps, such as yellow jackets, castes become distinct. In temperate regions and also in many tropical species, usually one queen is present per nest and she is considerably larger than her daughters, the workers. Colonies may be founded by a single queen or by swarming. In the former, the queen selects a nesting site, begins nest construction, forages, and cares for the young until they reach adulthood as workers. Swarming, a process in which a queen and a group of workers desert the nest, is common in the tropics and a new nest colony is formed rapidly by such a migrant group of wasps. Colonies founded by either a single queen or swarming soon number in the thousands and produce a large enclosed nest of papery material constructed from masticated wood mixed with salivary secretions. Nests start as a single layer of cells or comb, similar to those of primitive social wasps, but usually have many layers of multicelled combs added during expansion, each suspended under the preceding one by means of either stems or fastened to the

Figure 178. A new *Polistes fuscatus* nest founded by this single fertile female. The nest is open and manufactured from masticated wood fibers mixed with saliva.

Figure 179. The same *Polistes fuscatus* colony as on the previous figure two months later. Some of the cells contain eggs, others nearly mature larvae. Pupae are in the capped cells. Two central females are chewing up arthropod prey to be fed to the larvae.

lateral envelope or covering (Fig. 180). One or more entrances are constructed and guarded by workers (Fig. 181).

Similar to the termites, nests maintain a relatively uniform environment for the occupants, particularly the developing young. Covered nests shade larvae and pupae from direct exposure to sunlight and are layered to give insulation. Nests constructed in the ground vary the least in internal microclimate while the more common aerial nests become subject to greater climatic changes, particularly in the temperate regions. If temperatures rise too high, workers use their wings to create air flow to cool the nest.

Figure 180. A Polybiine wasp nest with part of the outer carton or covering removed to expose the inner cells. The white capped cells contain pupae. (Photograph by C. W. Rettenmeyer.)

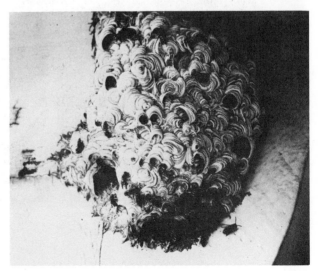

Figure 181. A paper wasp nest with many openings. Concentric rings are the result of different types of wood being added to the nest.

Social wasps feed their young daily. Food consists mainly of pellets of masticated prey (usually insects) or sometimes dead material; this diet differs from that of the adults whose diet is nectar, but some of the prey fluids may be swallowed as the food is being chewed for the larvae, and cannibalism may occur. In return for the food, workers often receive watery secretions from the larvae (trophallaxis) which probably aids in maintaining the social structure of the colony. After several weeks of growth, the larva spins a silken cap over the cell, pupates for from 3 to 4 weeks, and then emerges as an adult.

Ants

Over 12,000 species of insects are ants. All are social. Primitive ants are wasplike and exist in small colonies, often with only 2 or 3 dozen individuals. Little differences in polymorphism or polyethism exist between workers and queen. Eggs are not laid in cells, as wasp eggs are, but are deposited openly in the nest. After hatching, larvae are fed large pieces of unchewed arthropods, a feature unique to ants.

The most advanced ants have great diversity in the size and shapes of the worker caste. Although there is considerable overlap, the smaller forms (minor and media workers) tend to do more of the typical worker activities (Fig. 182) while the large ones (majors) generally defend the colony. Majors may have hard flattened heads to block entrances to the nest or long piercing mandibles for active defense (Fig. 183).

Figure 182. The size disparity between the large queen leaf-cutter ant (*Atta*) and the small minor workers. The white background is fungus, used for food, which grows on leaves brought back to the nest.

Figure 183. Army ant major (*Eciton burchelli*) with its ice tong type mandibles. Trophallaxis from workers is the only means of obtaining food for this defense specialist.

Reproductives have wings as they leave the nest to form swarms, in which mating occurs, with reproductives from other nests. Males then die while mated females lose their wings and seek out suitable areas for nesting. Once a chamber has been established, oviposition begins. In most higher species the hatched larvae are fed saliva from the queen and not prey as in primitive ants. Nutrients in the saliva and energy for the queen during this early period are derived from the reabsorption of her flight muscles and fat body. Fully grown larvae then pupate with the more primitive species producing a cocoon (Fig. 184) while the more specialized species have naked pupae.

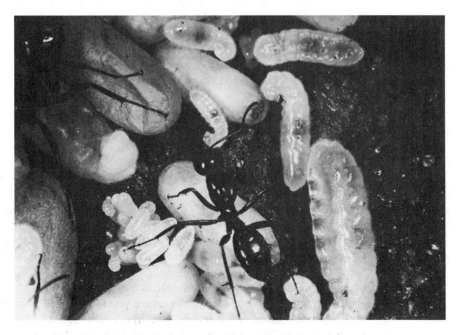

Figure 184. Eggs, larvae, enclosed pupae, and adult carpenter ant (*Campanotus pennsylvanicus*).

185

Figure 185. Worker carpenter ant (*Campanotus* sp.) in defensive pose. Formic acid can be sprayed forward at an antagonist.

Ant colonies, unlike most termites, increase rapidly in size during the first year. A colony may have one or several active queens. Queens may survive from 12 to 17 years and workers from 5 to 6 years. In some ants, when population size becomes excessive, the colony may divide, each with a separate queen.

A false impression that one receives from observing ants is the superficial busy atmosphere. In actuality, most individuals are inactive. If food is scarce, as many as 30 percent of the workers (Fig. 185) may be seeking food, but a "well-fed" colony may have as few as 5 percent searching at a given period (Sudd, 1967).

Ants have the widest range of diet of all social insects. Some are carnivorous, others are scavengers, while a few tend various species of Homoptera for sugar secretions. Many species have also turned to plants and their products for food. Seeds, fruit, and nectar from special plant structures, the *nectaries,* are commonly utilized. A few plants produce specialized fruitlike structures that specific ants use for food. Ants may subsist on fungi raised in specialized nest chambers, or *fungus gardens.* In the latter case, the fungus is specific and departing queens take spores with them in head pouches for initiating new fungal gardens.

Trophallaxis plays a greater role in ants than it does in the social wasps. Ant larvae are fed by adults and in return produce fluids, from the body skin and not from saliva as in wasps, that the adults ingest. This interaction is believed to provide a firm base for maintaining the functioning of the social roles within the colony.

Each ant often deposits material during its movements which results in "trails." These "trails" can be followed by others and are important in finding food, direction to nests, and often in determining distance.

Slave making, although uncommon, is found in ants. In one type, the queen enters another ant species nest and captures a portion of the brood. This slave brood adopts her and rears the new queen's offspring until the slave makers are able to assume all roles of colony survival. Another variation occurs when the invader or slave maker species has workers that are modified for fighting or where continual forays to nearby nests are necessary to obtain immature individuals that function as true workers after metamorphosis. Interestingly, the enslaved individuals adopt the new colony, will defend it, and feed the larvae and adults of the slave makers as if they were of the same species.

Nests of ants vary greatly but in no instance are they comparable to the highly evolved finite structures of the specialized termites, bees, and wasps. In harvester ants, nests consist of a gravel mound that provides good water drainage. Optimal temperatures for the occupants are obtained by vertical migrations. In the morning and evening, eggs and larvae are carried upward in the cone for warmth from solar radiation. During the extremely warm afternoon heat, the brood is transported downward to cooler sites. In the tropics, leaf-cutter ants may have massive nests of over 20 ft (6.1 m) in diameter that have multiple openings, and that descend over 10 ft (3.05 m) in depth. Some ants have become arboreal and the entire colony lives in hollow twigs or thorns, often resulting in protection for the plant. A few ant species such as the army ant, lack nests; the permanent nest is replaced with a temporary structure or *bivouac* (Fig. 186) formed by interconnecting legs and bodies. In the new

Figure 186. An army ant bivouac (*Eciton burchelli*) formed by the intertwining of ant bodies and legs. (Photograph by C. W. Rettenmeyer.)

world, this bivouac contains the brood and remains formed for less than 24 hours except during periods when the queen lays eggs.

Social Bees

Although having evolved independently at least eight times, social bees parallel the social wasps, especially in the temperate regions. In temperate climates, both normally start new colonies each spring from overwintering inseminated females (colonies in the tropics are usually perennial). Differences between queens and workers in primitive species are often only one of dominance behavior. Unlike the food for wasps, the food for bees consists of either nectar or pollen mixed with nectar. Also, certain brood cells are often used for storage of food (this is in contrast to the social wasps). The more conspicuous social bees are the bumblebees, honeybees, and stingless bees.

Bumblebees vary from most insects in being more numerous in the temperate than in the tropical region. An overwintering female starts the new nest in deserted rodent burrows or bird nests. Initially, a honey pot is fashioned from wax and is filled with nectar. Subsequently, brood cells are constructed, a few eggs are deposited, and the resulting young are fed pollen and/or nectar. Full-grown larvae spin cocoons and pupate for approximately 2 weeks. Adults that emerge are all females, usually smaller than their mother, and are subordinated by her aggressive behavior. These workers either stay in the colony and feed the young, convert the old brood cells to new storage tanks, or become foragers for nectar and pollen. Growth during the summer may result in several thousand individuals, but several hundred is more common. In large colonies, a few workers may also lay eggs but these become males or drones (parthenogenesis).

In the more advanced honeybees or stingless bees of the tropics, caste distinctions (Fig. 187) are more obvious. New colonies originate when a queen leaves with a swarm of up to 30,000 workers (queens cannot survive alone). Scouts locate a new site where a nest is started. New colonies grow rapidly, reaching approximately from 50,000 to 100,000.

Larval honeybees develop in wax cells (Fig. 188) where they are fed worker jelly, a secretion that has low amounts of hypopharyngeal gland material mixed with mandibular gland secretions and honey. After 3 days, hypopharyngeal secretions are withdrawn and beebread, honey mixed with pollen, is fed to the larvae. This switch in diet and a reduced number of feedings seem to produce stunted, sterile females, the workers. Worker development requires approximately 21 days. After emerging, honeybee workers pass through a series of behavioral steps during their short existence of approximately 6 weeks, progressing from no apparent work initially to feeding the young to producing wax (from special abdominal glands) to foraging for nectar and pollen.

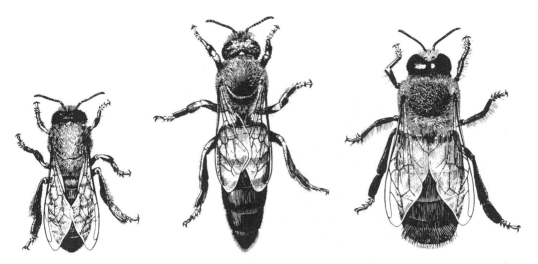

Figure 187. The honeybee castes (*Apis mellifera*). (A) worker or sterile female; (B) queen or fertile female; (C) drone or male, (*Courtesy* Agricultural Research Service, U.S.D.A.)

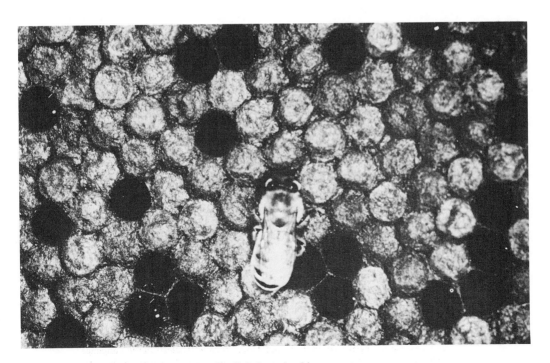

Figure 188. Capped brood cells in honeybee hive.

189

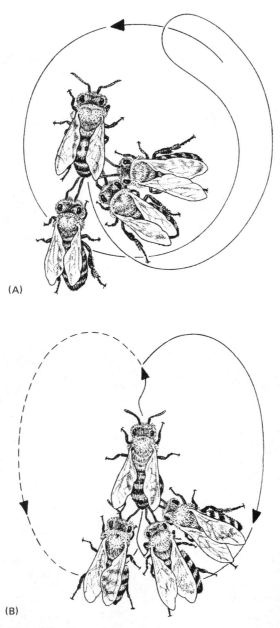

(A)

(B)

Figure 189. Types of dances by honeybees used in communicating nectar sources. (A) round dance indicating a source near the hive; (B) wagging dance indicating direction (angle of the straight run to gravity is same as angle of nectar to sun) and distance is farther away than in the round dance. (*From* "Dialects in the Language of the Bees" by Carl von Frisch, Copyright 1962 by Scientific American, Inc. All rights reserved.)

Normally, in higher bees, there is only one queen per colony because of at least two inhibiting pheromones in a material produced by the queen called the *"queen substance."* Following the death or departure of the old queen in a swarm, this inhibition is removed and the surviving workers are stimulated to transfer several eggs or newly hatched larvae to prepared queen cells. This is apparently to "remind" the workers to feed the female larvae in a different manner. Royal jelly, a substance that has high amounts of hypopharangeal secretions, must be fed continually to female larvae to produce queens. After approximately 16 days of development, the first queen emerges and the other queen pupae are systematically destroyed. If two queens emerge at the same time, a battle ensues until only one survives. She then takes a nuptial flight with drones, mates, and returns to propagate the colony. Approximately 1,000 eggs must be oviposited per day to maintain a large colony.

Honeybees, like ants, utilize the sun and polarized light for orientation during foraging activities. Once a food source is located, workers return to the colony and communicate the direction (in relation to the sun) and distance to the food by a series of dances (each species or strain of honeybee has slightly different dances and interpretations). Figure 189 illustrates two such dances. A strong, vigorous dance creates excitement because the food source is near and abundant; a weak dance elicits little response, and hive energy is therefore conserved. This communication by dances is unusual and isn't found in most bees, although some stingless bees may leave a scent trail to assist others in locating food.

Several hundred thousand flowers may be visited by a single bee each day to collect pollen or nectar. Usually, a bee forages for either nectar of pollen, but a number of worker bees (17 percent) take both on the same trip. Pollen adheres to the body hair and is then combed off and packed into special baskets on the outer surface of the hind tibia. Nectar is sucked into the crop, or "honey sac." When the bee returns to the hive, the pollen ball is scraped off each hind leg and is packed into a wax cell while nectar is either fed to others or regurgitated into cells for storage. Evaporation of water occurs and the honey is capped (Figs. 190, 202B). Normally from 15 to 100 lb. are needed as a reserve for the commercial colony (Fig. 191), depending on climate and colony size. Foraging flights for pollen or nectar requires considerable energy. For every continuous 2 hours or activity, each bee consumes its weight in sugar.

Bees in the tropics inhabit a uniformly warm environment with rain and predators acting as the major limiting factors. The honeybee in the temperate region, however, must regulate temperatures in the nest to counteract both heat of the summer and cold of winter. Warm temperatures (approximately 92 °F or 33 °C) are necessary for rapid brood development. When temperatures rise too high and are potentially detrimental for the colony, honeybees fan their wings (Fig. 192) and increase air flow and evaporation of water within the nest. During cold periods (below 50 °F or 10 °C), honeybees uniquely form a massive aggregation called a cluster. Metabolism, insulation produced by the

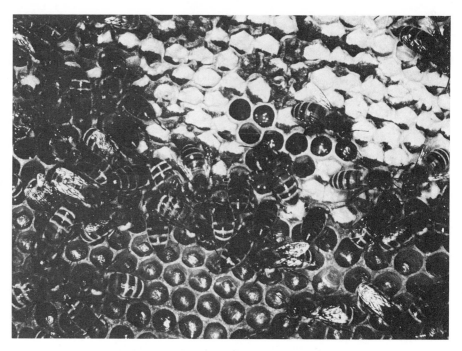

Figure 190. Honeybees (*Apis mellifera*) and honey production. Some cells are capped while others are in the process of being filled.

Figure 191. Honeybees (*Apis mellifera*) kept under commercial conditions.

192

Figure 192. Worker honeybees (*Apis mellifera*) fanning wings at hive entrance to cool the hive (ambient temperature over 95 °F).

layered hairy bodies, and alternate movement in and out of the cluster maintain heat far above the hive and ambient temperatures. Energy for this homeostatic system comes from the fat body and stored honey.

Questions

1. What criteria are used to define a social insect?
2. What are the advantages and disadvantages of each individual's being a member of a colony? Do these differ from the advantages and disadvantages gained by the colony as a whole?
3. What determines castes in termites and honeybees?
4. What selective pressures are exerted upon a colony as a result of individuals' having to undergo complete metamorphosis?
5. What means of defense do insect colonies possess?
6. What are inquilines? How might this relationship have arisen?
7. Are social insects better able to modify the environment than solitary species?
8. How are new colonies formed by termites, ants, honeybees, and social wasps?

8

Insects,
Plants,
and Humans

Humans are becoming aware that the increasing human population and the associated food production have both direct and indirect impacts upon ecosystems and that changes must be carefully evaluated as to their short-term and long-term effects. We, as humans, tend to view the components of nature and changes within the ecosystem as beneficial or detrimental to ourselves. Although this often creates problems, such as the erroneous belief that insects must be categorized as either good or bad, there are distinct advantages to be gained from such an approach and some of the major aspects will be discussed in this chapter.

BENEFICIAL ASPECTS

Pollination

More than 65 percent of the flowering plants (angiosperms) are pollinated by animals such as insects, birds, snails, and bats, but insects play the dominant role. The interrelationships between insect and flowering plant probably existed back in the Cretaceous period (over 125 million years ago). There is little doubt that each of these two groups has had a profound effect upon the evolution of the other. Current evidence, discussed by Baker (1968), indicates that most primitive angiosperms were pollinated by insects. Two major categories of plant adaptations that insure or increase the efficiency of insect pollination are those that (1) attract the insect to the flower so that pollen will be encountered and (2) those that "reward" the insect with either pollen or nectar.

Insects are attracted to flowers by different stimuli, and slight modifications result in specificity of this relationship. Petal colors of yellow, blue, blue–green, purple, or those that reflect or absorb high amounts of ultraviolet light are especially attractive. Nectar guides (Fig. 193B), pigmentation arrangement near the center of the flower, are often vital, as demonstrated by Daumer (1958). Daumer removed petals and reversed them so that the normal attachment points were toward the periphery. This rotation caused bee pollinators to move to the outside of the flower where the guides were located rather than to the center where they normally were. Flowers may also be located by smells, some of which are specific. An example of this is the odor

Figure 193. What an insect sees isn't necessarily what humans visualize. (A) photograph of daisy as man would see it; (B) the same flower to a bee. The inner parts of the petals, the invisible (to us) nectar guides, absorb ultraviolet light whereas the lateral portions reflect these wave lengths.

(A)

(B)

Figure 194.   Carrion flower (*Scapelia*) has a brownish-purple color and a rotting aroma, both of which attract such pollinators as blowflies and flesh flies.

that resembles decaying meat produced by certain flowers (Fig. 194) that attract only carrion-feeding insects. A few rare but interesting flowers resemble female wasps; pollen is picked up by the male wasp as it attempts to copulate with this mimic.

The second plant adaptation to insure insect pollination is that of "rewarding" insects that visit the flower. Historically, pollen, itself, most likely served as the initial benefit received by the insect, but nectar-producing structures, the nectaries (Fig. 195), have evolved and have greatly supplemented pollen. Plants gain through secreting an attractive nectar, but the nectar is sufficiently low in volume in order to prevent satiation of the insect. Early satiation reduces the number of flowers visited and, hence, pollination frequency drops. Since nectar is often located deep within the flower, only insects that have long probosci, such as in lepidoptera and bees, are able to feed (Figs. 196, 197).

Insect efficiency in utilizing flowers may be both structural and behavioral. First, the seeking behavior must occur during periods when flowering and nectar flow occurs. For generalized feeders, this may involve activity during several hours of most days, but for the more specialized insect, this also involves having life cycles synchronized with the plant. The orchid

Figure 195. Nectaries produce nectar that attract many insects. Most plants produce minimal amounts of the sugar solution, but the above poinsettia exudes copious amounts.

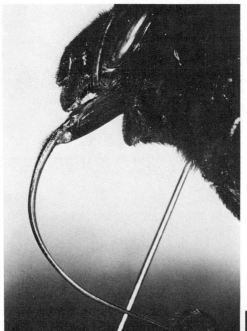

Figure 196. Orchid bee with elongated mouthparts for entering deep chambered orchids to obtain nectar.

Figure 197. Anthophorid bee taking nectar from snow-on-the-mountain (*Euphorbia* sp.) flower.

Figure 198.   Orchid bee (*Euglossa* sp.) with a pollinia attached. This pollen-bearing structure will be rubbed off on the next flower. (Photograph by C. W. Rettenmeyer.)

bees (Fig. 198) and yucca moths are interesting examples of specific pollination and the mutualistic relationship between insects and plants. The yucca moth gathers a pollen ball and carries it to another flower where it is forced onto the stigma. An egg is then layed on the flower and the resultant larva feed upon some of the developing seeds. Pollination, therefore, insures food for the caterpillar and surplus seeds allow plant propagation. Second, pollen and/or nectar must be collected and ingested efficiently. Pollen is often chewed from the anther by some insects such as beetles. Bees, however, illustrate a higher degree of specialization. As honeybees move about the flower, pollen is picked up by plumose body hair; the pollen is brushed off by their legs into pollen baskets on the hind legs. Insects that feed on nectar usually have chewing–lapping, siphoning, or sponging mouthparts although many exceptions are evident.

The co-evolution of flowering plants and insects has resulted in a multitude of species for each group. Insect orders that have numerous species (Coleoptera, Lepidoptera, Hymenoptera, Diptera) have been associated with flowering plants for thousands of years. Of the plants pollinated by these insects, many are important to our economy and include beans, peas, tomatoes, most fruit, cotton, alfalfa, tea, and cocoa. In a few instances the interrelationships have become obligatory, for neither species can reproduce without the other (e.g., figs and fig wasp, orchids and orchid bees) although such relationships are potentially hazardous to both if populations of either partner become reduced in number.

Trashburners and Soil Builders

The word *trashburner* has been used by many biologists to indicate organisms involved in degradation and recycling dead plants and animals. This process is particularly important in preventing a permanent tie-up of necessary nutrients. The most common examples of such decomposers are the bacteria and fungi; however, many insects feed directly on and burrow into both plant and animal cadavers (Fig. 199). The insects not only directly assist in the recycling but also break the outer barriers of the cadaver and enhance invasion by bacteria and fungi. In dry areas a dead animal untouched by multicellular burrowers or scavengers will mummify long before much decay can occur. Further evidence of the importance of insects in the recycling process may be seen in Australia where an estimated 25 percent of all wood is decomposed by the feeding of termites.

Soil building is closely related to decomposition. The breakdown of both plant and animal remains and the mixing of this material with soil aids in developing layers of humus. The role of earthworms in soil building is widely

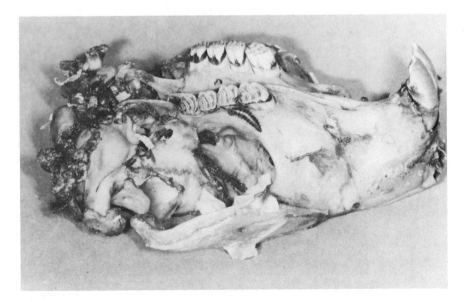

Figure 199.   Dermestid beetle larvae feeding on dried flesh of beaver skull. Museum mammalogists often use dermestid beetle larvae to clean skulls.

recognized. Although no complete studies comparing earthworms with insect activity in soil formation and renovation are known by the author, undoubtedly the role of insects is significant and, according to a few scientists, must at least approach the effectiveness of earthworms in some areas of the world, for example, the tropics. Preliminary information from studies of termites and soil formation in Australia (Lee and Wood, 1971) indicate that these insects are unique in several ways. First, instead of turning over the A horizon as earthworms do, they selectively bring up wet soil particles from the B horizon to incorporate them into their nests. Second, such nutrients such as calcium, magnesium, and potassium are often concentrated in and around the colonies because of their feeding activities and these become tied up for long periods because the nests do not decompose readily.

Products

Silk.   One of the few "domesticated" insects is the silkworm moth, *Bombyx mori* (Fig. 200), for it has become essentially dependent on humans for survival. This moth is reared mainly in the Orient, the Mediterranean, and South America. Although rearing once was restricted to small huts and buildings, modern technology now enables mass breeding of silkworms in factories. Here

Figure 200.   Silkworm adults (*Bombyx mori*) mating. The male was attracted to the female by sex pheromones released from everted lobes near her genitalia. They immediately coupled and remained *in copula* for over three days.

Figure 201. Commercial silk production now involves mass production. Killing the pupae is vital to prevent adult metamorphosis and emergence that would damage the cocoon. (A) silkworm cocoons moved by a conveyer from drying oven to packaging area; (B) cocoons of silk worms opened and compared. The one on the left is before entering the oven while that on the right demonstrates the results of oven exposure to the pupae.

eggs are processed under sanitary conditions and workers entering rearing rooms must comply with strict regulations of clothing and personal cleanliness. After hatching, larvae are transferred to rearing trays, and mulberry leaves, the only food of preference, are added until the maximal 3-inch size is reached by the caterpillars. Within a week, the cocoon is placed in a high-temperature drying oven that kills the pupa (Fig. 201). The silken cocoons are then bagged and delivered to a processing plant, often thousands of miles distant, where they are unwound and processed into commercial thread and cloth. Approximately one ton of mulberry leaves are thus transformed into 12 lb. of silk.

Honey. Honey, the product of the honeybee, *Apis mellifera,* represents concentrated and partly hydrolyzed nectar. The principal sugar in nectar is sucrose which is broken into glucose and fructose by salivary enzymes as the bee returns to the hive. This fluid is regurgitated from the crop into comb cells where the solution is concentrated by bees fanning their wings and evaporating water from the mixture. Once the mixture is concentrated to approximately 80 percent sugar, the cells are sealed with wax for future use by the bees although humans usually intervene at this stage and reap the rewards.

Most areas in the United States have at least 100 different species of plants that are visited by honeybees for nectar, but only a small fraction are used extensively. Specific tasting honey can be produced by placing hives near appropriate nectar sources since each source has its own aroma and influence on the taste of the product. Alfalfa, citrus, cotton, and clover are the most common sources of honey.

To harvest honey, special hives are maintained. A series of separate movable sections, each containing vertical wax combs, or *supers,* are stacked one on top of another. These supers are inspected periodically to detemine the time of harvesting (Fig. 202). People usually wear protective clothing to prevent being stung; one never knows when hypersensitivity to bee stings may develop and result in anaphylactic shock and death unless promptly treated. Once the combs are full and the honey is sealed by the bees, supers are removed and the honey is processed.

Figure 202.   Honey production using manufactured nests of stacked supers. (A) supers are periodically removed and the combs checked for the stage of honey production; (B) comb in the process of being filled with cells uncapped.

Beeswax.   Honeybees produce beeswax to build their combs. About the twelfth day as an adult worker, specialized hypodermal glands on the fourth to seventh abdominal segments start to secrete thin flakes, which are scraped off by the hind legs and molded by the mandibles. Energy comes from metabolizing honey and it is estimated that from 3 lb. to 20 lb. are requried to produce 1 lb. of wax. In 1969 in the United States, over 5½ million lb. were harvested by beekeepers for over $3 million. Beeswax is utilized in innumerable products from such items as smokeless candles and carbon paper to cosmetics.

Other Substances.   Additional materials synthesized by insects and used in our society include royal jelly (from honeybees), cantheridin (from certain blister beetles), and lac (from scale insects). Further readings in this area are available in Metcalf, Flint, and Metcalf (1951).

Food

The realization that birds were predaceous on insects is ancient. According to some studies (McAtee, 1932; Reed, 1943), from 50 to 60 percent of the food of birds are insects. The actual number of insects eaten varies with the bird species; Reed (1943) estimated that 100 ingested per day is average, but flickers have been found to have over 5,000 ants in their stomachs. Although most species utilize this source of food only during their juvenile development, a large number of birds including warblers, woodpeckers, swallows, and vireos continue to feed upon insects throughout their life. If insect populations decline, birds migrate or alter their diet. Meadowlarks, for example, feed on these arthropods primarily during the spring and summer but eat seeds and other food in cold weather. Many persons, perhaps unintentionally, have attempted to persuade the uninformed that birds should be protected because of their destruction of pest species; however, the role of avian predators has been greatly exaggerated except in isolated situations. Normally, birds do not specialize solely on pest species and their impact upon such populations is only minimal.

Mammals include insects in their diet. Some are occasional consumers, such as bears, while others, including moles, bats, and shrews, have become nearly specific insect feeders. Control recommendations for moles, for example, often include soil treatment for beetle grubs, the reasoning being that once the primary source of food is eliminated these insectivores either starve or migrate to other areas. Most bats feed exclusively on flying insects and locate them by echo location. Bat cries bounce off the prey, and position is determined by the strength and time of the returning sound vibrations. Certain noctuid moths have specialized sense structures, tympana on the thorax, that hear the high-pitched bat cries and the moths immediately initiate a series of evasive flying maneuvers. Typical "herbivores" such as deer mice also feed extensively on insects.

Figure 203. Insects are important components in the diet of game fish. This bull-head has captured a large cockroach, found it to be positioned incorrectly to swallow, and ejected it. Moments later the roach was taken in correctly, with the insect's head first, and swallowed.

Predation by cold-blooded vertebrates is extensive although qualitative and quantitative data is incomplete when compared to birds and mammals. Insects make up more than 75 percent of the diet of the common toad, *Bufo* sp. Lizards such as *Sceloporus* sp., chamelions, horned toads, and green glass snakes are effective insectivores. An estimated 40 percent to 90 percent of the diet of freshwater fish is of insect origin (Fig. 203). Trout, for example, ingest great quantities including mayflies, stoneflies, and caddisflies. One fish species, *Gambusia affinis*, has been used extensively in controlling mosquito larvae in city parks and extensive marshes where other control measures are impractical.

Although largely ignored by the "civilized" western world, insects are a good source of food for humans. Many areas of the world have and still derive part of their protein from insect sources. The American Indian, for example, often included recipes for insect preparation in his culture; immatures of the common brine fly, *Ephydra hians*, and adults of *Atherix* flies have been harvested extensively in California. Grasshoppers are eaten by many cultures worldwide as are termites and large caterpillars. Most of these insects are roasted or boiled, but some are eaten uncooked. Insects contain few parasites of humans, are of high food value (Table 2), and are readily obtained. A change in our thinking is needed in order to use insects to feed a portion of the growing human population but this change is not anticipated in the near future.

TABLE 2.   Comparative Value of Selected Insects and Other Foods

Food	Water	Percent Protein	Fat	CHO	Ash	Calories of 100 g
Living termites	44.5	23.2	28.3	—	—	347
Silkworm pupae*	60.7	23.1	14.2	0.5	—	207
Grasshoppers*	70.6	18.7	4.1	—	—	—
Turkey	64.2	20.1	14.7	0	1.0	218
Beef (standard)	60.0	18.0	21.0	0	0.9	260
Cod fish	81.2	17.6	0.3	0	1.2	78
Crab	78.5	17.3	1.6	0.5	1.8	93
Lobster	78.5	16.9	1.9	0.5	2.2	91
Eggs, hen	73.7	12.8	11.5	0.9	1.0	163
Clams	81.7	12.6	1.6	0.5	1.8	76
Milk, cow's	87.4	3.5	3.5	4.9	0.7	65

*Some of the data are not available for comparative purposes.

Control of Other Organisms

Fifteen of the 26 orders of insects have species that naturally parasitize or prey on other organisms that humans classify as detrimental. Hymenoptera and Diptera contain the greatest number. Our deliberate use of these insects, as well as our use of other organisms such as bacteria in management programs, is termed *biological control,* but we have much to learn in this area.

One of the best examples of successful biological control by insects occurred in Australia. By 1920 over 20 species of the cactus *Opuntia* had invaded and become sufficiently established over 60 million acres to make many rangelands nearly useless. Mechanical and chemical control proved unsuccessful in preventing their spread. Of the 50 species of cactus-feeding insects tested, 12 indicated sufficient promise to be introduced, but only the lepidopteran caterpillar, *Cactoblastis cactorum,* and several cochineal mealybug species proved effective. Their success was astounding, and several of the cacti species disappeared with the remainder now found only in sparse, isolated populations.

Scientific Experiments

Until recently the use of insects for scientific experiments has not been widespread because of their small size and attendant requirement for delicate and precise instrumentation. Genetics has been a discipline in which some species have been used extensively. Several fruitfly species (*Drosophila*) have been valuable test animals. Much of our current knowledge of mutations, sex-linked inheritance, and other classical genetic studies was obtained from experiments with these flies.

Population dynamics is another area that utilizes insects. Research on the migratory locust and many of the stored product pests has added much to our understanding of density-dependent factors upon population growth.

With more refined instrumentation the scientific community will continue to turn to insects as experimental animals because of their rapid life cycle, relative structural and behavioral simplicity, low cost of rearing, and diversity.

DETRIMENTAL ASPECTS

Human Disease Transmission

Although this relationship was not realized until the last 60 to 70 years, insects serve as *vectors* (carriers) of the major human diseases listed in Table 3 as well as certain enteric diseases of lesser concern. Most of these major diseases have had a profound effect upon humans and their civilizations. Some excellent examples include the death of approximately 25 percent of all Europeans in 1348 because of the plague (Table 4); the ending of Napolean's dreams of conquest because of typhus (and the Russian winters); the delaying of the completion of the Panama Canal because of malaria and yellow fever (they cost the French alone some $200 million and 50,000 lives); and the exclusion of humans from thousands of acres of land in Africa because of African sleeping sickness.

Malaria (Fig. 204) has been and is still considered one of the major world health problems, especially during wars. Until the attempt at worldwide eradication (since 1947), which brought the number of cases down to nearly 100 million per year, from 5 percent to 15 percent of the world's population were expected to contract this disease each year. The number of cases in the United States has also dropped, from an estimated 6 million cases in 1935 to 129 cases in 1968. One of the leading factors in this decline has been the widespread use of residual insecticides, especially DDT; however, because of the concern over the long life of these chemicals and the cost of using substitute insecticides and antimalarial drugs, humans have apparently reached their capacity to control this disease until some further breakthrough occurs. In fact, concern is being expressed about an increased incidence of malaria in the United States, for approximately 4,000 import cases were reported in 1971, and such cases could serve to infect the endemic mosquito vectors.

Several advantages for pathogens are gained through associations with insects including:

1. selectivity and locating the host by the vector
2. protection of the pathogens during the transmission phase

TABLE 3.   Major Human Diseases Transmitted by Insects*

Disease	Causative Organism	Carried by Certain	Order
Human malaria	Protozoa	Mosquitoes	Diptera
Yellow fever	Virus	Mosquitoes	Diptera
Dengue	Virus	Mosquitoes	Diptera
Encephalitis, Arborvirus Groups A and B	Virus	Mosquitoes	Diptera
Filariasis	Nematode	Mosquitoes	Diptera
Onchocerciasis	Nematode	Black flies	Diptera
Tularemia	Bacteria	Deer flies†	Diptera
African sleeping sickness	Protozoa	Tsetse fly	Diptera
Bubonic plague	Bacteria	Fleas	Siphonaptera
Murine typhus	Rickettsia	Fleas†	Siphonaptera
Epidemic typhus	Rickettsia	Human lice	Anoplura
Chagas disease	Protozoa	Assassin bugs	Hemiptera

*Specific names of pathogens and vectors are omitted.
†Causative organisms may be carried by other vectors.

TABLE 4.   Major Epidemics of Bubonic Plague*

Country or City	Year	Deaths
Western Europe	1348	25,000,000†
London	1603	33,347
London	1625	41,313
Messina	1656	48,000†
London	1665	68,596
Marseilles	1720	88,000
Moscow	1771	60,000†
Egypt	1834	32,000
Canton	1894	40-100,000†
Punjab	1906	675,307

*Since complete records of most epidemics are unreliable, only a few of the more reliable records are listed. Compare only to the relatively low populations of the date and not with current levels. Also, much of the widespread transmission of this disease was through secondary pneumonic action rather than fleas.
†Estimated

3. relatively rapid dispersal
4. utilization of the insect for multiplication purposes
5. inoculation through the protective skin barrier of the host

Insects normally do not serve as long-term reservoirs because of their short life. Mechanical transfer of pathogens also may occur by carrying them on some portion of the body, such as the mouthparts and feet.

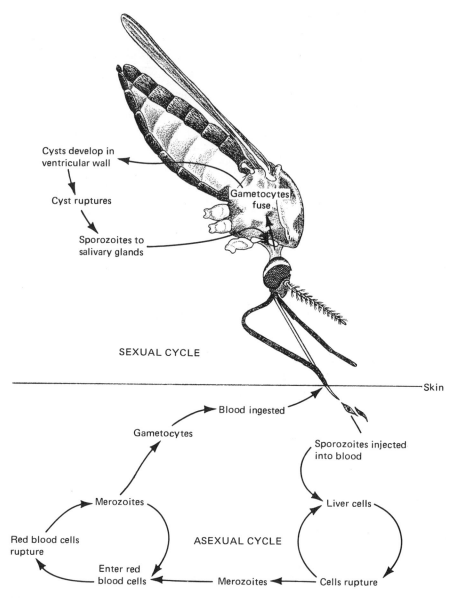

Figure 204. Life cycle of *Plasmodium* sp., the causative protozoan of human malaria. Sexual phase is in the mosquito, the asexual in man.

Stings, Bites, and Allergies

Many insects either sting or bite humans. In most instances, these stings and bites only annoy (Fig. 205), but allergic responses and confirmed deaths do occur each year. In the years 1950–1959, nearly twice the number of people died in the United States from the stings of hymenopterous insects (229) than from poisonous snakes (139). Those who died from these stings usually did so

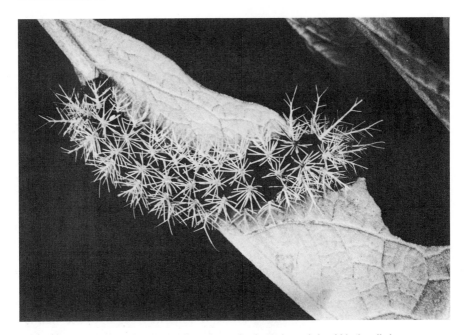

Figure 205.   This tropical *Automeris* larva has urticating hairs and should be handled with care.

within the first hour and probably were allergic from previous exposures. Although these figures should not produce panic, they do nevertheless, represent factual occurences and should invoke precautions when dealing with insects.

In addition to Hymenoptera, the following insects commonly bite man: Diptera (mosquitoes, black flies, stable flies, horseflies, deer flies, tsetse flies, sand flies); Hemiptera (bed bugs, assassin bugs, backswimmers, giant water bugs, nabid bugs); Siphonaptera (fleas); and Anoplura (sucking lice). Since biting or stinging insects are normally encountered outdoors on picnics or hikes, simple precautions such as insect repellents and proper clothing will markedly reduce the incidence.

In addition to stings and bites, numerous allergic responses have been reported. These include some respiratory reactions, commonly diagnosed as hay fever and asthma. Some scientists have gone so far as to state that insects are a major cause of many of these respiratory allergies.

Entomophobia

In some individuals, simple annoyance by insects and arthropods grows into psychotic behavior termed *entomophobia* or *delusory parasitosis*. Victims of entomophobia imagine that insects are jumping at them, are feeding on them, are making loud noises while they are walking around at night, are ruining their food, crawling on their skin, and so forth. Distressed individuals are in earnest about their "observations" and require immediate psychiatric aid, but it is normally difficult, at best, to convince these people that they need help. Pomerantz (1959) and Waldron (1962) review a few typical case histories.

Livestock

Insects are frequent parasites of domestic animals. Lice are parasitic in all post-ovarian stages and remain on the host during their entire life cycle. Other species spend only a developmental stage upon the host, usually as larvae, e.g., bot flies. Some feed only as adults, e.g., biting flies. Especially in high populations, parasitic insects cause irritability, produce decreased eating by the host, and more susceptibility to other diseases. Weakened animals are of low market value or, in the case of dairy cows, yield reduced milk quantities.

In addition, insects transmit over 50 livestock diseases. An excellent example exists in the epidemic of Venezuelan equine encephalitis (VEE) in 1971. Over 1,500 horses died of this mosquito-borne disease in Texas during a massive campaign to isolate and control this introduced disease from Central and South America. Only after hundreds of thousands of acres were aerially sprayed, over 3 million horses in 19 states were vaccinated with a relatively untested serum, and quarantine restrictions were placed on movement of horses was this serious viral disease controlled. In addition to disease transmission, insects also serve as intermediate hosts of parasitic worms (Table 5).

TABLE 5. Examples of Parasites in Which Insects Serve as Intermediate Hosts (Note the Wide Divergence of Insects Involved)

Parasite	Causative Organism	Host	Insect Intermediate
Haematoloechus medioplevus	Fluke	Frogs	Dragonfly nymphs
Crepidostomium cooperi	Fluke	Fish	Mayfly nymphs
Prosthogonimus macrorchis	Fluke	Birds	Dragonfly nymphs
Dipylidium carinum	Tapeworm	Dogs, cats	Fleas, lice, beetles
Hymenolopis carioca	Tapeworm	Chickens	Beetles, house flies
Hymenolopis fraterna	Tapeworm	Rats, mice	Beetles
Moniliformis dubius	Acanthocephalid	Pigs	Cockroaches
Diplotriacnoides translucida	Nematode	Birds	Grasshoppers
Skrjabinoptera phrynosoma	Nematode	Lizards	Ants
Dirofilaria immitis	Nematode	Dogs	Mosquitoes, biting flies
Habronema megastoma	Nematode	Horses	House flies

Plants

Previous discussions have indicated the long associations of flowering plants and insects. Many of these relationships have been beneficial, as indicated with pollination, but many have been harmful. It is estimated that approximately 50 percent of the insect species, especially during the immature stages, use living plant material for food. It seems ironic that the larva may feed upon the plant and be considered detrimental, but the adult, because of its pollination activities, is considered beneficial. Selectivity may be so specific that only a given species, genus, or family of plant is fed upon. In contrast, generalists (polyphagous) feed upon a wide variety of plants and are subsequently less restricted. But as insects have become more efficient herbivores, plants that are capable of withstanding the insect pressures also increase in numbers.

Nearly all plant structures are eaten by some insect (Figs. 206, 207, 208, 209). If feeding is extensive, the plant may die, but most plants are able to withstand insect infestations under normal conditions, at least to propagate. What defenses do plants have? (1) Plants may regenerate lost structures at a fast rate. (2) Some plants have structures including spines, pubescens, leaf shape, and thick cuticles that physically interfere with insect feeding.

Figure 206.   Some larvae feed only on the outer layers of leaves, a process termed skeletonizing. The photograph illustrates extreme skeletonizing of elm leaves by the elm leaf beetle *(Pyrrhalta luteola)*.

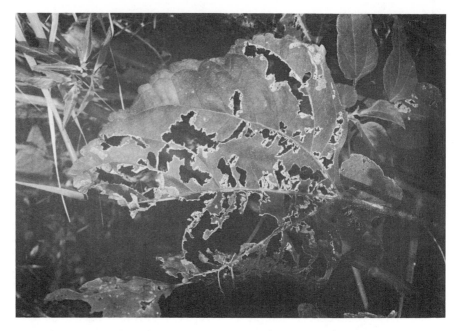

Figure 207. The above damage on sunflowers occurs when the insect eats the entire cross section of the leaf.

Figure 208. This southwestern corn borer (*Diatraea grandiosella*) has severely damaged the corn plant causing it to lodge under high winds. The larva overwinters in the base of the corn stalk.

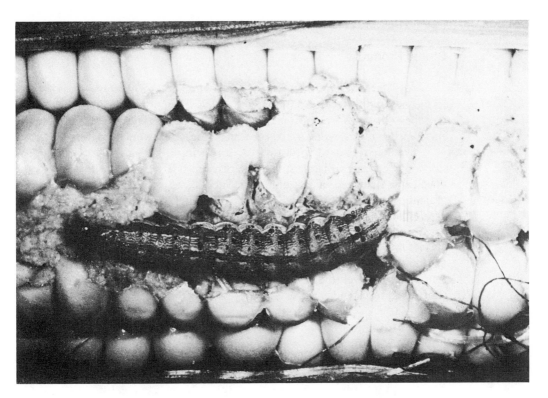

Figure 209. The larva feeding on this corn is the corn earworm (*Heliothis zea*). A voracious feeder, this pest is very repulsive to consumers.

(3) Relocation of certain vital nutrients for insect survival may occur. Nitrogen and starch are often in low percentages in old leaves, especially in climax vegetation. (4) Timing of growth so that plants that develop early and withstand the effects of nature may escape the full attack of insects whose appearance is synchronized with the main plant population. (5) Wide dispersal decreases the likelihood of insects locating individual plants. Some scientists believe this to be a major factor in plant distribution in the tropics. (6) Secondary plant substances often discourage some insects from feeding or ovipositing. Tannins, resins, and silicone are common substances that interfere with normal insect feeding, especially plants with K-strategy (climax vegetation). Steroids, pyrethrins, nicotine, and saponins are often toxic to herbivores that do not specialize upon plants that produce these compounds.

Damage to the plants may come from other than feeding. Insects may carry plant disease, primarily viral and microplasmal diseases. The difficulty of a virus disseminating itself from a sessile plant is obvious. Insects, therefore, have become a ready means of distribution because of their intimacy and host species preferences, especially those that have piercing–sucking mouthparts and ingest sap containing the disease agent. Over 250 viruses are known to be transmitted by insects, with the majority having species of Homoptera as vectors; aphids alone transmit over 160 of these.

Stored Products

Any harvested product, whether it is of animal or plant origin, is subject to the attack of insects. Common materials fed upon include grains (Fig. 210) and their products, seeds, nuts, fruits and vegetables, meats, dairy products, furs, leather goods, wood and its products (Fig. 211), wool, and tobacco.

Damage to stored foodstuffs by direct feeding of insects or contamination may be great, with an estimated 5 percent to 10 percent of the world's produce being damaged or lost annually. Such losses are equal to the amount of grain required for the survival of approximately 130 million people. Losses are greatest in the tropics where food, especially grains, cannot be adequately dried and hence massive infestations of both insect (Fig. 210) and fungal species may be seen. To prevent this, some grains such as corn are left in the field until used. When stored, grain must be maintained at or below the 12 percent moisture level to deter infestations, but fumigation and/or spraying with short-lived insecticides are usually necessary. Of major concern in the United

Figure 210. Granary weevils (*Sitophilus granarius*) can cause considerable damage to stored grain left undisturbed for some length of time. Eggs are laid in cavities of corn kernels where the larvae often complete development within four to seven weeks.

Figure 211.   This book damage is the result of termite feeding.

States are approximately 50 species belonging primarily to the order Coleoptera and Lepidoptera. The important beetle species produce damage during both the adult and larval stages, but moths do so only as larvae. A knowledge of the development and ecology of each insect pest is needed before specific infestation problems can be solved.

INSECT CONTROL

Approximately one-third of the world's human food supply is being destroyed by pests while millions of people are starving. Humans, in their attempts to raise more food as well as protect themselves from biting and disease-carrying insects, have devised methods to alter normal population growth of many species of insects. Early practices were to minimize damage but, with the advent of synthesized insecticides in the 1940s, attempts became directed toward eradication. Despite these efforts, successful eradication of a single pest insect species is unknown except from such localized areas as islands. In the late 1960s and early 1970s a new era began when, as the result of concern for the environment, *pest management* emerged as a dominant concept, i.e., regulating populations instead of attempting to eliminate them. Killing 75 percent of a pest population at critical times, in many instances, may be sufficient to prevent economic damage while maintaining predators and parasites at sufficient levels to permit efficient long-term regulation. A discussion of the major ways of managing pests is in order. Of prime importance in selecting one or more of these methods into an *integrated management program* are the following questions: (1) What is economic damage? (2) During what stage in plant development is damage critical? (3) What is the value of the crop per acre? (4) What manipulations achieve the best results? (5) What will be the effect of control measures on other potential pests, predators, and the ecosystem.

Legal and Cultural Control

Restrictions upon people and their manufactured products often play an important role in limiting the distribution of potential insect pests and the diseases they may carry. In a few instances, quarantines have restricted serious buildups and spread of exotic or introduced pest species. A good example may be seen when Venezuelan encephalitis, a serious viral disease of horses in Central and South America, appeared in Texas in 1971. By restricting interstate shipment of horses in the southwest, this mosquito-borne disease was limited, and an intense program of mosquito control and horse immunization was employed with high effectiveness. Even with these stringent measures, over 1,500 horses died in one month.

Legislation is having an increasing effect upon control practices. At the federal level, the Federal Insecticide Act, passed in 1910, was designed to protect consumers against the fraudulent selling of misbranded or adulterated products. Years went by with no significant legislation until 1947 when the Federal Insecticide, Fungicide and Rodenticide Act was passed which transferred to the manufacturer the responsibility of proving insecticide effectiveness; this effectiveness was to be demonstrated by adequate research before the approval and release of the product onto the market. The Miller Amendment (1954) to the Federal Food, Drug, and Cosmetic Act introduced the safety factor of residues. Subsequent concern over carcinogenic effects have entered the picture. Because of these legislative changes, even more intensive screening by companies and university experiment stations for insecticidal effectiveness, dosage requirements, and safety factors must precede certification. In December 1970 the Environmental Protection Agency was created, and in 1972, through the Federal Environmental Pesticide Control Act, the agency was charged with the regulation of all pesticide usage, including certification of the product and applicators using certain restricted insecticides. State and local agencies have generally been concerned with the local distribution, storage, and spraying recommendations. Land-grant universities have been in the foreground here and nearly all publish yearly recommendations for the use of insecticides (Fig. 212) on the crops in their areas.

Cultural techniques are vital in managing insect pests. Rotation of crops, soil tillage at proper times, and timing planting and harvest often mean the difference between an average or good crop.

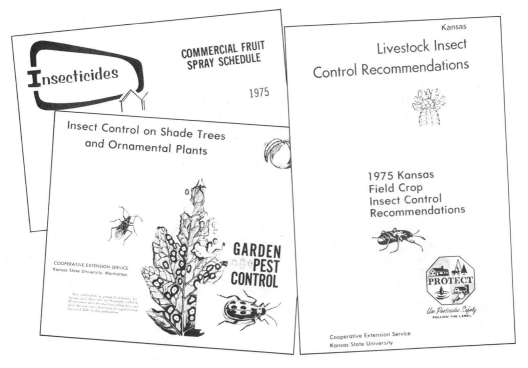

Figure 212. Many states regularly publish recommendations for insect control. This photograph includes a few examples of those distributed in Kansas.

Some varieties of crops are more resistant to pests than others (Figs. 213, 214). Painter (1968) categorized plant resistance into three major types:

1. *tolerance,* or the ability to survive high pest infestation that would severely injure or kill susceptible plants
2. *nonpreferred,* when the plant seems to be "ignored" by pests
3. *antibiosis,* the ability to induce detrimental effects on the pest and its development

Breeding for resistance to pests, especially when coupled with other control measures, has achieved good results with certain crops. The Hessian fly (*Mayetiola destructor*), which nearly wiped out wheat in some midwest areas several decades ago, can now be controlled by combining resistant varieties of wheat with late planting. Because of the current concern over pesticide use, this area of pest management is now receiving much attention.

Biological Control

Insects, as previously indicated, are food for a wide spectrum of organisms including other insects (Figs. 215, 216, 217, 218), bacteria and fungi (Fig. 219), vertebrates, and plants (Fig. 220). Under certain circumstances, in-

217

Figure 213. Plant breeding experiments on wheat indicating those plants resistant (left), tolerant (middle), and susceptible (right) to Hessian fly infestations.

Figure 214. Experiments to select alfalfa plant strains resistant to pea aphid attacks. Those on the left are much more susceptible than those on the right as determined by height and lateral growth. (Photograph by E. L. Sorensen.)

Figure 215. This aphid infestation on milo is being brought under control by wasp parasites. Note the white mummies, which are the exoskeletal remains of parasitized aphids.

Figure 216. Carolina mantid (*Stagmomantis carolina*) feeding on a tree cricket (*Oecanthus niveus*).

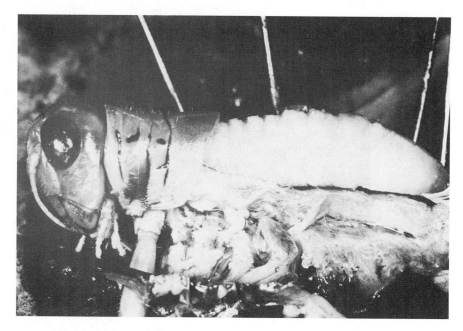

Figure 217. A short-horned grasshopper (*Melanoplus differentialis*), dissected, revealing dorsally a larval Nemestrinid fly. Air to the parasite is obtained by attaching to the host's tracheal system.

Figure 218. This sphingid larva will not complete development because of Brachonid wasp parasites. These wasps, after feeding within the host, bore out and pupate in silken cocoons as seen in the photograph below.

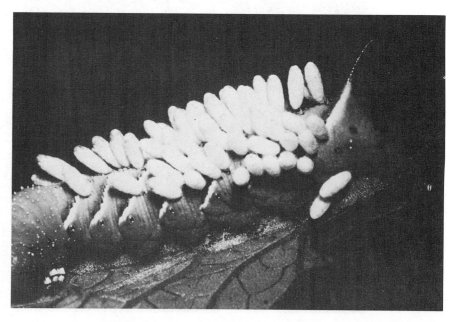

Figure 219. Biological control of insect pests has proved to be an alternative for use of chemicals in certain instances. In some circumstances the environment is "seeded" with a specific pathogen of the pest. The above *H. cecropia* larva, although not normally considered a pest, is dying from a fungus that has caused extreme dehydration through diarrhea, a situation of lethal consequence in caterpillars, which must maintain a turgid condition to locomote.

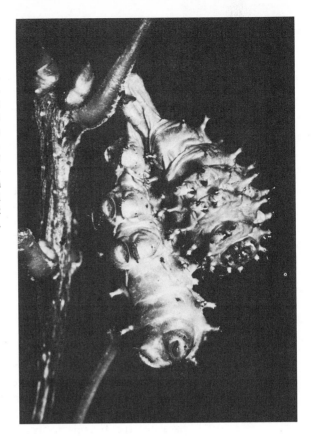

Figure 220. A few insects, such as this house fly (*Musca domestica*), fall victims to plant predation. This venus flytrap has just reopened, exposing the fly remains.

sect pests may be controlled by the introduction of these pathogens and predators. Biological control programs require considerable scientific information on population potentials and fluctuations of both pest and control agents, possible ecological ramifications, expected variations from locality to locality, and practical economics. The following requirements must be satisfied for effective long-term pest management:

1. there must be a close synchronization of host and parasite cycles
2. predators and parasites must be effective during the growth phase of the host population
3. host populations must be controlled before economic damage
4. predators or parasites must be mass reared for release when native populations lag behind the host
5. the climate must be conducive to predator survival
6. alternative hosts or a residual of pest individuals must be left to maintain predators unless continual mass releases are contemplated.

The advantage in introducing parasites and carnivores instead of insecticides is that the ecosystem is less affected.

Several successful pest management programs have been reported, including the control of the Mediterranean fruitfly in Hawaii by an insect parasite, the European spruce sawfly in Canada by a virus, the cottony cushion scale in the United States by an insect predator, and the Japanese beetle by bacteria. The Japanese beetle is still a problem, for only short-term results have been achieved.

Genetic Control

Insect management by mass sterilization of males and releasing them to compete for mates with native males is another potential technique. The screwworm fly control program in Florida and the southwest United States serves as the best example known. Screwworm fly larvae have cost the cattle industry an estimated $100 million each year until the mid-1960s. Observations that the number of matings of female flies is limited, often to a single event, suggested success through male sterilization, particularly since this species is restricted geographically to a warm winter habitat. Initially, these flies were successfully eliminated from Florida and concentration was shifted to Texas where a facility capable of rearing 200 million sterile flies per week (Bushland, 1975) has been constructed. Most of the southwest has become nearly free of these pests and with the completion of a facility in Mexico capable of producing 300 million sterile flies per week, permanent control is hoped for. Programs such as the screwworm fly are costly (approximately $5 million for the Texas project); therefore, the pest must be of high economic significance or a potential health hazard to warrant such concerted expenditures.

Hormones

If there is a means of selectively controlling insects, it would logically lie in the manipulation of some factor that insects require for survival or reproducing. Pheromones appear to have such control implications, especially when combined with insecticides and traps. The strategy is to attract specific pests to a centralized area, thereby reducing the need to "blanket" spray areas and kill other insects, including carnivores and parasites. An example is the synthetic compound Disparlure which is a sex attractant for male gypsy moths, a severe pest of forests in eastern United States. Although the potentials of Disparlure are still being investigated, preliminary experiments indicate success for isolated areas or regions of new infestation. Control possibilities using this technique seem promising.

Since insects require juvenile hormones during their developments, what are the possibilities of manipulating concentrations during critical periods? Can pest species and not other insects be affected in a locality? A species of European Hemiptera could not be reared in United States laboratories because the paper toweling contained a juvenile hormone mimic. Other examples of juvenile hormones and synthetic analogs are known to have been effective in limited testing trials. The discovery of insect anti-juvenile hormones in plants (Bowers et al., 1976), substances that interfere with the titer of juvenile hormones, may prove to be monumental. Therefore, the possibility of utilizing hormones in insect control does exist, and much testing is currently being conducted to refine these techniques. It is doubtful, however, that this method will be a panacea.

Mechanical and Physical Control

Except for screens on windows, light traps, and certain instances of handpicking, this control probably has the most limited potential of those discussed. Microwaves and other radio-frequency energy have been used experimentally to kill some pests, for example, those on stored grain, but their cost on large-scale operations is currently prohibitive.

Insecticides

Insecticides are compounds that are designed to kill insects. Each insecticide must be registered. Registering involves much scientific research on their reliability, ability to "selectively" kill insects, safety to humans, and rates of use. It takes approximately 5 years and it is expensive for the manufacturers. Over 200 insecticides are registered, but only a few are used extensively (Table 6). It should be noted that although these appear to selectively kill insects, they also have a varying effect upon other animals, especially when (1) the concentrations are increased by improper formulation of sprays, (2) continual exposure because of carelessness, (3) buildups in soil because of poor drainage

TABLE 6.   Some Commonly Used Insecticides—Present and Past

Botanicals	Inorganics	Synthetic Organics
Nicotine	Arsenates	Aldrin*
Pyrethrin	Fluorides	Chlordane*
Rotenone	Paris green	DDT*
	Sulfur	Diazinon†
		Dieldrin*
		Heptachlor*
		Malathion†
		Methoxychlor*
		Parathion*
		Sevin‡
		Toxaphene†

*Chlorinated hydrocarbon compounds.
†Organic Phosphate compounds.
‡Carbamate compounds.

and the use of long-lived residuals, (4) runoffs from sprayed areas into streams, and (5) the concentrating effect of some food chains as higher trophic levels are reached.

An insecticide can be used one year and achieve its result, but over 95 percent of a population of insect pests must be killed to have a decrease in next year's numbers. Since normally, this isn't possible, yearly spray schedules are instituted. Insect pests that survive this regimen are genetically better adapted for withstanding subsequent application, and thus resistance develops. Also, other animal species that survive this treatment increase abnormally in numbers because of less predation and competition, and sometimes they become more detrimental than the original pest, at least during the agriecosystem recovery period.

Insecticides are often grouped into *botanicals, inorganics,* and *synthetic organics.* Of the botanicals, tobacco was used in France as early as 1763, the major active agent being nicotine, which comprised approximately from 2 percent to 5 percent of the leaves. Pyrethrins, a group of naturally occcurring botanical insecticides, was used in the early 1800s in Asia. Both nicotine and pyrethrins are contact poisons and are available on the market today. Another contact poison from plants, rotenone, was first used about 1850.

The use of botanical insecticides was complemented by inorganic compounds in the late 1800s and early 1900s; inorganic compounds were far more toxic, but they had to be ingested to kill. First paris green and then the arsenates, fluorides, sulfur, and so forth were introduced, but no controls were placed on them until 1926 when their lethal potential to humans, animals, and plants alike were finally considered hazardous. Until the discovery of DDT, paris green was the major larvacide in mosquito control, and, undoubtedly, it aided in the decline of malaria in the United States. Arsenates are still available and, although used only in certain specific situations, these compounds remain among the major causes of pesticidal poisonings in children.

DDT, the first of the new synthetic organic insecticides, was introduced extensively at the end of World War II. Although synthesized in 1874 and proven to have insect-killing qualities in 1936, this chlorinated hydrocarbon required the seriousness of malaria and epidemic typhus in World War II to bring about its experimentation and introduction. Its effective action, long life, and apparent safety were immediately hailed as the possible end to our insect problems. Soon it was used extensively. The impossible seemed probable and such things as eradication of malaria and safe spraying of crops appeared just over the horizon. Over 65,000 tons were used worldwide in 1962, the peak year of its use. Resistance of many insects to its action and the detection of increasing residue concentrations all over the earth have reversed public opinion, however.

It is estimated that approximately 75 percent of the total land area in the United States has never been treated by insecticides. Nevertheless, misuse, accidental poisonings and contaminations, the number of inexperienced users of these chemicals (15 percent are homeowners), lack of scientific information, and many other reasons have created a re-evaluation in the philosophy and use of insecticides. Apparent findings of DDT residues in regions of the world where application have never been made, such as in Antarctica, lend fuel to this fire. It is safe to say that the current trend of substituting the organic phosphates and carbamid compounds for the long-residual hydrocarbons will continue (Fig. 221).

Figure 221. Spraying can be done using minimal equipment when more sophisticated machinery isn't available. This make-shift sprayer has been put together by enterprising farmers in South America. (Photograph by T. A. Granovsky.)

Figure 222. Elm tree killed by a fungus transmitted by bark beetles. Elms killed by this Dutch Elm disease die suddenly with the dead leaves often retained. Because of the frequency with which elm roots graft when meeting, this disease may also pass through the plant vascular system from one to another.

Figure 223. One method of preventing Dutch Elm disease is through systemic insecticides. This elm has been measured and insecticide is entering the tree through plastic tubes at a rate calculated from the trunk diameter. Most uses of systemics, however, are through soil or foliage applications, (Photograph by H. E. Thompson.)

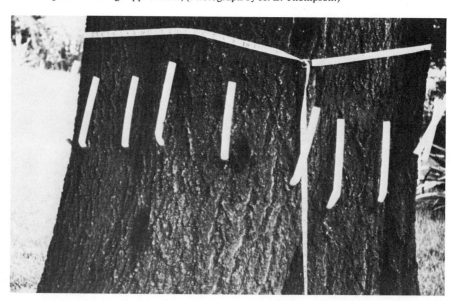

Alternatives to wide-spread insecticide use are being investigated, especially in a concept of *integrated control* (using several methods to obtain maximal results). These include biological control by releasing predators or pathogens of the pest at appropriate intervals, mass rearing and sterilization of males of certain insect pests (only possible in some flies at the present time) that mate only once and releasing them in high number to compete with native populations, the use of sex attractants to lure pests into traps or insecticides, the production of plants and animals that resist the action of the pest, the use of juvenile hormones to prevent maturation of the young and subsequent reproduction, injection of the insecticide into the plant (Figs. 222, 223), as well as others. Nevertheless, at least in the near future, insecticides will be a necessary part of management programs.

Decisions now facing the world include whether or not we, as a dominant force in the biosphere, can permit further buildups of persistent insecticidal residues even though their qualities and low dollar cost have resulted in near eradication of malaria and other insect-borne diseases in some areas and an increase in the food supply for the ever increasing human population? Can we monetarily afford to use the relatively high priced, short-lived but more immediately toxic substitutes that are even more vulnerable to misuse? Are we prepared to live in an environment without insecticide use? Can we effectively use these compounds in a limited way until other means are developed or in conjunction with other methods in an integrated management program? Humans, indeed, have some complicated and far-reaching decisions to make.

Questions

1. What structural and behavioral modifications do insects have to benefit from plant associations? What advantages and disadvantages have plants gained from this relationship?

2. Do you think that humans will utilize insects for food in the near future? Why or why not?

3. How are insects beneficial to humans? How are they detrimental?

4. What insect orders are of major economic and medical concern to humans?

5. Compare the advantages and disadvantages of each of the major types of control methods. Do you feel that a combination of several techniques is better than utilizing one type? Explain.

9
Classification

One has to make only a cursory look at nature to conclude that there are many types and degrees of resemblances and differences in organisms. Many systems have been devised to put tags or names on these groups of plants and animals. The use of common names has been extensive and has the advantage of easy communication because common names often illustrate obvious biological peculiarities, but these names often indicate different organisms depending on the geographical locality, i.e., potato bugs can refer either to a crustacean or several species of insects.

Out of the many diverse naming systems came the one used today, which has as its basis the 1735 work of Linnaeus, *Systema naturae*. This early enumeration included over 4,000 species of animals. One hundred years later Agassiz and Bronn counted 129,370. The estimate today varies but is usually over 900,000 kinds with approximately 75 percent of these species classified as insects (Fig. 6).

The naming of this great assemblage pioneered by Linnaeus has the scientific name as a binomial (consisting of two parts). The first part indicates the *genus* and the second the *specific*. The adoption of this procedure has resulted in a standardization of names throughout the world; for example, *Musca domestica* has the same meaning in Russia as it does in Japan or Brazil. As a further refinement, species have been assembled into a classification hierarchy of similarities, a process referred to as *classification*. This hierarchy can be illustrated as follows for the house fly, *Musca domestica:*

Kingdom	Animalia
Phylum	Arthropoda
Class	Insecta
Order	Diptera
Family	Muscidae
Genus	*Musca*
Specific	*domestica*

Closely related or similar species are grouped into the same genus (pl = genera). Similar genera are placed within the same family. This procedure is carried on up the hierarchy until the classification is complete. The common families, orders, and so forth have well-established common names and these also have been used throughout the text. A knowledge of the classification system is important, for it is vital to data interpretation, storage, and retrieval.

Various keys have been formulated to assist in identification. Usually these consist of a series of numbered couplets that have two choices. After deciding which choice is correct for the given specimen, proceed to the proper couplet as directed by the right-hand number. By observing carefully, answering each choice correctly, and proceeding through the key an unknown insect can be identified and classified according to the various levels within the hierarchy. For introductory students, most keys are restricted to the higher levels of the classification such as the phyla, classes, and the major orders and families.

The following two keys should prove useful as an introduction to the major insect orders commonly found in the United States. Such uncommon orders as Protura, Entotrophi (Diplura), Embioptera, and Zoraptera are omitted. Keys to families within the larger orders deal only with those normally seen by introductory students. More detailed keys, i.e., including more diverse and less common species, are available in advanced taxonomic texts.

Key to the Orders of Adult Insects
Normally Found in Insect Collections

1. Wings present. 12
   Wings absent 2
2. With long cerci (Fig. 224E). . . . . THYSANURA
   Without long cerci (Fig. 244) 3
3. Large insects, usually 1 in. (2.54 cm) or more in length. . . . . ORTHOPTERA
   Small insects, usually ½ in. (1.27 cm) or less in length 4

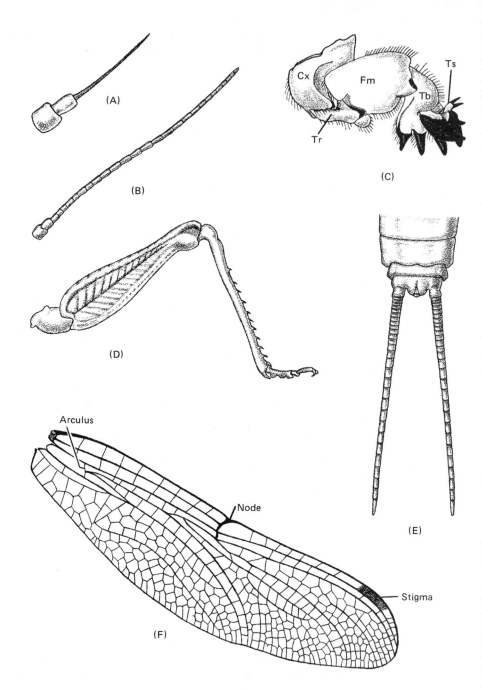

Figure 224. Diagnostic characters used in identifying adult insects. (A) setaceous antenna (Ephemeroptera); (B) filiform antenna (Orthoptera); (C) digging leg (Orthoptera); (D) jumping leg (Orthoptera); (E) long cerci (Plecoptera); (F) node, stigma, and arculus of wing (Odonata).

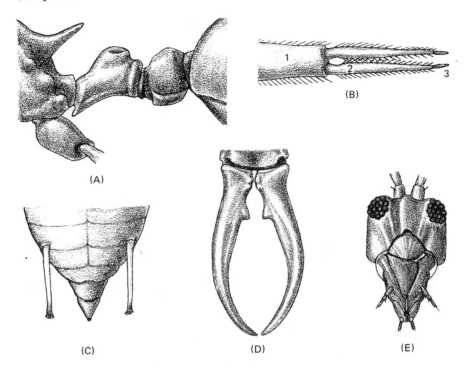

Figure 225. Diagnostic characters used in identifying adult insects. (A) slender waist (Hymenoptera); (B) ventral view of furcula (Collembola); (C) cornicles present on abdomen (Homoptera); (D) cerci forceplike (Dermaptera); (E) conelike beak (Thysanoptera).

4. Usually possess abdominal forked spring (furcula) (Figs. 225B, 228); minute size. . . . . COLLEMBOLA
    Small insects, usually ½ (1.27 cm) or less in length 5

5. Cornicles present (Fig. 225C). . . . . HOMOPTERA
    Cornicles absent 6

6. Abdomen constricted into narrow waist at juncture with thorax
    (Fig. 225A) . . . . HYMENOPTERA
    Abdomen not constricted into narrow waist 7

7. Rasping–sucking mouthparts in conelike beak (Fig. 225E); large bulging eyes; body long and slender, often posteriorly pointed
    (Fig. 247). . . . . THYSANOPTERA
    Chewing (Fig. 244) or piercing-sucking mouthparts (Fig. 252);
    eyes usually reduced (Fig. 246) 8

8. Antennae moniliform (Fig. 37D); short cerci present (Fig. 241) . . . . . ISOPTERA
    Antennae not moniliform; cerci absent (Fig. 245). 9

9. Antennae long and slender (Fig. 244) . . . . PSOCOPTERA
    Antennae short (Fig. 246) 10

10. Legs modified for jumping (Fig. 287); body flattened
     laterally .... SIPHONAPTERA
     Legs not modified for jumping; body flattened dorso-ventrally 11

11. Head usually broader than long (Fig. 245); chewing
     mouthparts.................................... MALLOPHAGA
     Head usually longer than broad (Fig. 246); piercing–sucking mouthparts,
     retracted into head .... ANOPLURA

12. Long cerci present (Fig. 230)................................ 13
     Cerci short (Fig. 241) or absent................................ 17

13. Antennae setaceous (Fig. 224A).................... EPHEMEROPTERA
     Antennae filiform (Fig. 224B)...................................... 14

14. Front pair of legs adapted for digging (Fig. 224C) or grasping
     (Fig. 240B).................................... ORTHOPTERA
     Front pair of legs similar to middle pair (Fig. 243A)..................... 15

15. Hind pair of legs enlarged for jumping (Figs. 224D, 238)...... ORTHOPTERA
     Hind pair of legs similar to middle pair (Fig. 243A) 16

16. With large unsegmented forceplike cerci (Fig. 225D)......... DERMAPTERA
     Cerci long and filamentous, many segmented (Fig. 243A)...... PLECOPTERA

17. With an anterior node or notch on each wing (Fig. 224F); antennae setaceous
     (Fig. 224A) .... ODONATA
     Without an anterior node or notch on each wing....................... 18

18. One pair of wings present; halteres present (Fig. 295A) .... DIPTERA
     Two pairs of wings present; halteres absent 19

19. Piercing–sucking mouthparts in elongate beak (Fig. 250) 20
     Mouthparts other than above................................ 22

20. Hind leg without tarsal claws, adapted for swimming (Fig. 50D).. HEMIPTERA
     Hind leg with tarsal claws................................ 21

21. Beak arises at anterior part of head (Fig. 252B); hemelytra usually present
     (Fig. 251A) .... HEMIPTERA
     Beak appears to arise between front pair of legs (Fig. 254)...... HOMOPTERA

22. Rasping–sucking mouthparts in conelike beak (Fig. 225E); wings fringed with
     long hair (Fig. 247) .... THYSANOPTERA
     Not as above................................ 23

23. Front pair of wings horny or of different texture from hind wings 24
     All wings membranous................................ 25

24. Front pair of wings thickened and hard, normally without veins
     (Fig. 266) .... COLEOPTERA
     Front pair of wings with veins; hind legs often enlarged for jumping
     (Fig. 238) .... ORTHOPTERA

25. All wings equal in size (Fig. 241) .... ISOPTERA
     Hind wings smaller than front pair of wings........................ 26

26. Siphoning mouthparts coiled under head (Fig. 283); wings and body usually
     covered with scales................................ LEPIDOPTERA
     No mouthparts coiled under head; scales absent or few in number, primarily on
     wing veins................................ 27

27. Many crossveins in wings, particularly at anterior edge
    (Fig. 260) . . . . NEUROPTERA
    Few crossveins in wings (Fig. 279). 28

28. Mouthparts reduced, only palpi obvious (Fig. 279) . . . . TRICHOPTERA
    Mouthparts not reduced 29

29. Chewing mouthparts, elongated and beaklike (Fig. 278). . . . . MECOPTERA
    Chewing mouthparts not elongated *or* chewing–lapping mouthparts
    (Fig. 196) . . . . HYMENOPTERA

Key to the Most Common Immature
Insects Found in Insect Collections

1. External wing pads present (Fig. 226A); compound eyes present
   (Fig. 227C) . . . . (nymphs) 12
   External wing pads absent; compound eyes absent (Fig. 226F). . . . . (larvae) 2

2. Body lacking legs (Fig. 226B) 3
   Body with legs present 6

3. Head present (may lack pigmentation) 4
   Without distinct head; (Fig. 226B); 1 pair large posterior spiracles usually
   present. . . . . DIPTERA

4. Body uniformly elongate (Fig. 226C) *or* terminal breathing tube present
   (Fig. 226D) . . . . DIPTERA
   Not as above. 5

5. One pair of thoracic spiracles (Fig. 267A). . . . . COLEOPTERA
   Two pairs of thoracic spiracles . . . . HYMENOPTERA

6. Abdominal prolegs present (Fig. 226G) 7
   Abdominal prolegs absent 8

7. Five or fewer pairs of abdominal prolegs (Fig. 226G) . . . . LEPIDOPTERA
   More than five pairs of abdominal prolegs (Fig. 274) . . . . HYMENOPTERA

8. Gills present on abdomen (Fig. 227A). 10
   Gills absent on abdomen 9

9. Body C-shaped (Fig. 267C); nota absent. . . . . COLEOPTERA
   Body not C-shaped or some nota present 11

10. Distinct lateral appendages on abdomen (Fig. 227A) . . . . NEUROPTERA
    Distinct lateral appendages on abdomen absent although thread-like gills may be
    present; larvae often in cases (Fig. 280) . . . . TRICHOPTERA

11. Abdomen without extensive sclerites (Fig. 262B) . . . . NEUROPTERA
    Abdomen with extensive hard sclerites (Fig. 267B) . . . . COLEOPTERA

12. Piercing–sucking mouthparts (Fig. 252B). 13
    Chewing mouthparts 14

13. Mouthparts arising at the anterior of the head (Fig. 252B);
    filiform antennae. . . . . HEMIPTERA
    Mouthparts appearing to arise near base of prothoracic legs; setaceous antennae
    often present . . . . HOMOPTERA

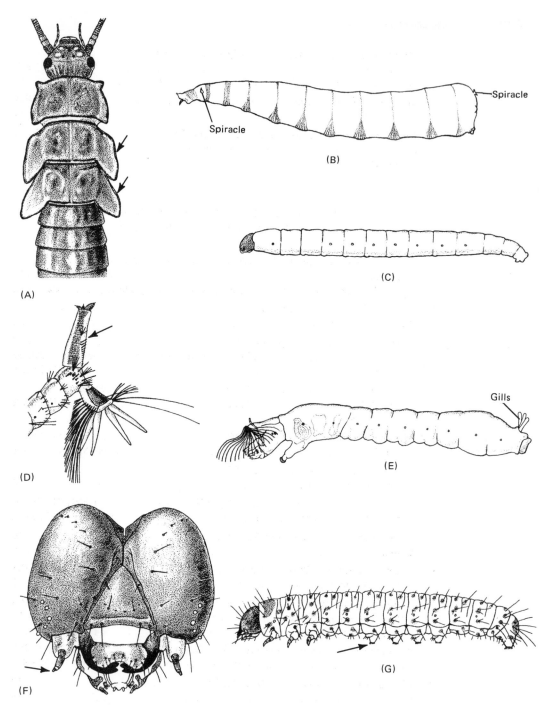

Figure 226. Diagnostic characters used in identifying immature insects. (A) external wing pads (Plecoptera); (B) distinct head absent (Diptera); (C) body uniformly elongate (Diptera); (D) terminal breathing tube (Diptera); (E) mouth fans (Diptera); (F) antennae short (Lepidoptera); (G) five pairs of abdominal prolegs (Lepidoptera).

234

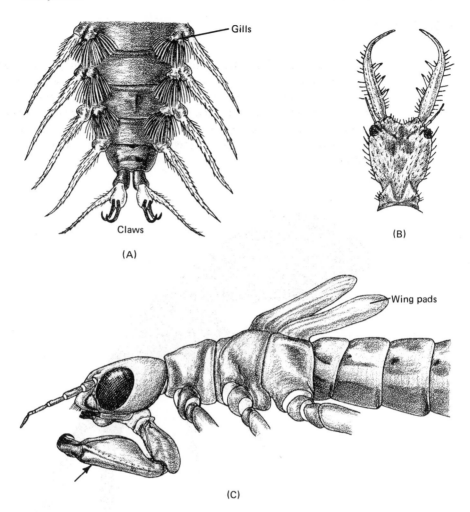

Figure 227. Diagnostic characters used in identifying immature insects. (A) lateral abdominal appendages, ventral view (Neuroptera); (B) mouthparts sickle-shaped and hollow (Neuroptera); (C) labium greatly enlarged but not extended, wing pads present (Odonata).

14. Labium longer than head when extended (Fig. 227C) . . . . ODONATA
    Labium not extendable, normal size 15

15. Three "tails" at tip of abdomen; abdominal gills present
    (Fig. 231) . . . . EPHEMEROPTERA
    Two "tails" or none at tip of abdomen (Fig. 243B). 16

16. Cerci long, ½ or more length of abdomen (Fig. 243B). . . . . PLECOPTERA
    Cerci short, less than ½ length of abdomen (Fig. 42) . . . . ORTHOPTERA

COLLEMBOLA

Springtails (Fig. 228) are minute, seldom exceeding 6 mm in length. They are easily recognized by their size, absence of wings, six-segmented abdomen, ventral tube, or *collophore,* on first abdominal segment, and the usual presence of an abdominal forked *furcula,* or spring; the latter, when present, is used to propel the organism forward. Springtail antennae are normally four-segmented and their eyes are represented by lateral groups of from one to eight stemmata.

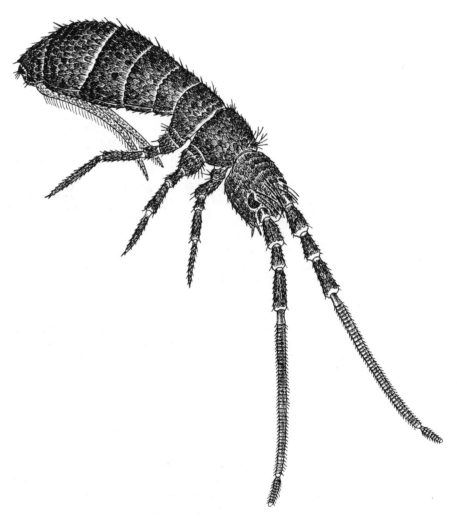

Figure 228.   A springtail.

Ametabolous development—except for size, all instars are similar in shape and, since external genitalia are absent, diagnosis of adults is based on whether or not the individuals reproduce. Molting occurs often after adulthood is reached, a trait unique for insects.

Five families have been proposed to include the approximately 2,000 species (approximately 200 in North America). Springtails are normally abundant in soil or leaf litter and feed mainly as scavengers or on algae. They become pests only rarely, such as the lucern flea on alfalfa, a Puerto Rican species transmitting disease of sugar cane, and damage to mushrooms, house and greenhouse plant seedlings.

THYSANURA

Bristletails range in size from 1 mm to 50 mm in length. They are wingless, have elongated filiform antennae, two well-developed cerci and a single caudal filament, and possess *styli* or vestigial legs on their abdomen. Mouthparts are of a weak chewing type, weak compound eyes are present, and the body may be covered with scales (Fig. 229).

Figure 229.   A silverfish (*Lepisma saccharina*).

Metamorphosis is lacking and the young resemble the adults. These insects move rapidly and some are capable of jumping. The 400 species (30 in North America) are divided into 4 families, one of which includes several pests to man. The firebrat *(Thermobia domestica)* is found near furnaces and steam pipes. The silverfish (*Lepisma saccharina*) (Fig. 229) seeks cooler areas that have higher moisture. Bookbindings, wallpaper paste, flour, and organic debris form the diet of these two pests.

EPHEMEROPTERA

Mayflies (Fig. 230) are from 4 mm to 50 mm in length and may be identified by their setaceous antennae and long multisegmented cerci. A median caudal filament may be present. Two pairs of wings are usually present (hind pair usually reduced and sometimes lacking), have numerous veins, and are normally held vertically over the thorax at rest (Fig. 230). Compound eyes are large and may be divided in males. Mouthparts are vestigial.

Metamorphosis is incomplete. Eggs are deposited in water. Nymphs (Fig. 231) have abdominal gills and ordinarily have three "tails," two pairs of cerci and a caudal filament. Some nymphs cling to rocks and are flattened and streamlined while others are cylindrical and burrow (Fig. 231) or move about on the bottom. Specific habitat preferences are common. Food consists of aquatic organisms, usually plant, and organic debris. After completing their immature development (1 to 2 years), nymphs leave the water and molt into a *subimago,* a winged immature stage unique to mayflies. Within hours to several days, another molt occurs and the adult emerges. Mating, often in

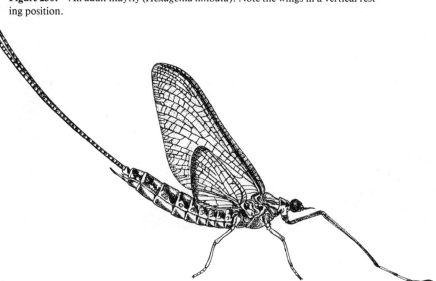

Figure 230.   An adult mayfly (*Hexagenia limbata*). Note the wings in a vertical resting position.

Figure 231. Mayfly nymphs are aquatic. This nymph (*Hexagenia limbata*) uses its tusklike mandibles for digging in the mud.

swarms, takes place and eggs are oviposited. Survival by adults is short as indicated by the derivation of the order name (*ephemerous* = living but a day).

Approximately 2,100 species are known (600 in North America) and these are categorized into 15 families. This group forms an important source of food for some fish but is of no direct economic or medical importance to humans.

ODONATA

These insects (Fig. 232) are medium to large (wingspread from 25 mm to 190 mm) and have the following characteristics: large compound eyes, strong chewing mouthparts, setaceous antennae, elongate slender abdomen with small single segmented cerci, and two pairs of long narrow wings with many veins, a notch, and an arculus (Fig. 224F). Two suborders exist in the United States. One, the Anisoptera, or dragonflies, hold their wings horizontally when at rest, have the thickest bodies, and are very active fliers whereas the Zygoptera, or damselflies, hold their wings nearly vertical (Fig. 233) when at rest, have very slender bodies, and are less agile in flight.

Males have secondary copulatory organs on the second abdominal segment. During copulation the male grasps the female's "neck" with his claspers while she curves her abdomen up to his second segment. Females then deposit

Figure 232. Examples of Odonata. (A) darner dragonfly (Aeshnidae); (B) broad-winged damselfly (Calopterygidae); (C) common skimmer (Libellulidae); (D) broad-winged damselfly (Calopterygidae).

Figure 233. An adult damselfly (*Agrion maculatum*).

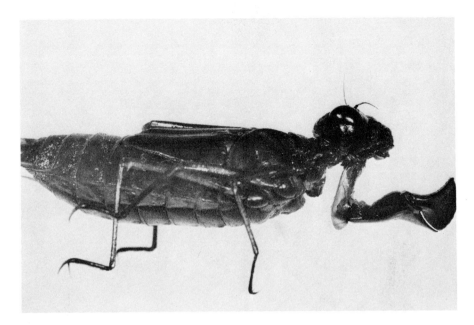

Figure 234. A dragonfly nymph. Note the enlarged labium and the external wing-pads.

eggs in or near water. Metamorphosis is incomplete. Nymphs are voracious predators and utilize an enlarged labium for capturing prey (Figs. 108, 234). Damselfly nymphs have three terminal platelike gills while oxygen exchange in dragonfly nymphs takes place internally in a musculated rectum. If the rectal muscles, used normally in exchanging water, contract strongly, water will be forced out rapidly and a form of "jet propulsion" occurs. After 3 to 4 years, nymphs crawl up emergent vegetation and metamorphose into adults.

The nearly 5,000 species (over 400 in North America) are separated into 9 families. Their value to humans is difficult to assess, for they are general feeders and, although many mosquitoes and other insects detrimental to humans are consumed, many beneficial species are also eaten.

ORTHOPTERA

This group (Fig. 235 includes such insects as cockroaches (Fig. 76), mantids (Fig. 236), walking sticks (Fig. 237), grasshoppers (Fig. 238), and crickets (Fig. 239). All possess chewing mouthparts, long legs with one- to five-segmented tarsi, and large compound eyes. Wings are usually present and have many veins and are modified with the forewings often narrowed and thickened into a *tegmen* while the hindwings are broad, membranous, and folded fanwise under the mesothoracic pair (Fig. 45). Flight is mainly through action of the hindwings.

Figure 235. Examples of common Orthoptera. (A) pigmy locust (Tetrigidae); (B) short-horned grasshopper (Acrididae); (C) cockroach (Blattidae); (D) mantis (Mantidae); (E) mole cricket (Gryllotalpidae; (F) cricket (Gryllidae); (G) long-horned grasshopper (Tettigoniidae); (H) camel cricket (Gryllacrididae); (I) walking stick (Phasmidae).

Figure 236. The Carolina mantis (*Stagomantis carolina*).

Figure 237. A walking stick (*Diapherimera persimilis*).

Figure 238. A long-horned grasshopper (*Scudderia texensis*).

Figure 239. A field cricket (*Acheta assimilis*).

Stridulation or sound production by scraping (Fig. 132) is a means of attracting mates. An appendicular ovipositor is common and often measures as long as the abdomen. Cerci are often short. Antennae commonly are elongate and multisegmented. Size ranges from 12 mm to over 250 mm in length.

Egg laying is variable. Some eggs are deposited in the soil (short-horned grasshoppers), but others are deposited in or on vegetation (long-horned grasshoppers). Walking sticks deposit eggs singly, but mantids and roaches produce an oötheca. Metamorphosis is incomplete. Most Orthoptera are herbivorous, but some are carnivorous (mantids) or omnivorous (cockroaches).

Some species are of economic importance. A few short-horned grasshoppers have been pests of crops throughout recorded history, especially in the temperate and arid regions of the world, and many areas in the United States conduct annual surveys to anticipate population increases that might reach economic thresholds. Migrating forms are of especial importance when populations are high. In some years the Mormon cricket, a wingless long-horned grasshopper of the intermountain west, has been very destructive because of its large size, gregariousness, and migratory behavior. Field crickets may damage seedlings in truck crops. Also, some species of cockroaches inhabit houses and become important pests.

Over 22,500 species exist worldwide with more than 1,000 in North America. These species are placed in 12 to 19 families, depending on the classification system used. Some systematists split this order into the Orthoptera and the Dictyoptera, the latter order including mantids, walking sticks, and cockroaches. The following key will help the beginning student to identify the more common families in the United States.

Key to Common Families of Orthoptera

1. Pronotum extending backward almost to tip of abdomen or sometimes beyond tip of abdomen (Fig. 240A) . . . . (pygmy locusts) TETRIGIDAE
    Pronotum not extending behind thorax (Figs. 240D, 240E) 2

2. Front tibiae and femora greatly enlarged for digging, with large black toothlike processes (Fig. 224C) . . . . (mole crickets) GRYLLOTALPIDAE
    Front tibiae and femora not greatly enlarged for digging 3

3. Hind femora much longer and thicker than middle femora (Fig. 238) 4
    Hind femora about the same length and thickness as middle femora. 7

4. Tarsi 3-segmented (Fig. 240C) 5
    Tarsi 4-segmented 6

5. Antennae about as long as pronotum
    . . . . (short-horned grasshoppers) ACRIDIDAE
    Antennae much longer than pronotum (Fig. 235F) . . . . . (crickets) GRYLLIDAE

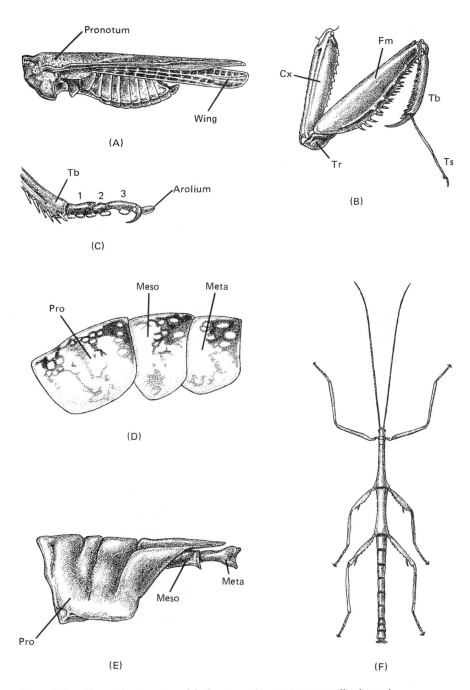

Figure 240. Diagnostic characters of Orthoptera. (A) pronotum extending beyond tip of abdomen (Tetrigidae); (B) grasping leg (Mantidae); (C) tarsus 3-segmented (Acrididae); (D) thoracic nota similar (Gryllacrididae); (E) thoracic nota dissimilar (Acrididae); (F) body sticklike (Phasmidae).

6. Wings absent or vestigial; mesonotum and metanotum similar to pronotum
    (Fig. 240D); prosternal spines absent
    . . . . (cave or camel crickets, sand crickets) GRYLLACRIDIDAE
    Wings large; mesonotum and metanotum very different from pronotum
    (Fig. 240E); prosternal spines usually present
    . . . . (long-horned grasshoppers) TETTIGONIIDAE
7. Front legs adapted for grasping prey (Fig. 240B); large spines on front tibiae and
    femora . . . . (mantids) MANTIDAE
    Front legs not adapted for grasping prey 8
8. Body narrow, elongate, sticklike (Fig. 240F) . . . . (walking sticks) PHASMIDAE
    Body broad and flat (Fig. 235C). . . . . (cockroaches) BLATTIDAE

ISOPTERA

Termites vary from 2 mm to 12 mm in length except for physogastric
queens. They are characterized by a prognathic head, moniliform antennae
with from 9 to 30 segments, chewing mouthparts, short and stout legs with
four-segmented tarsi normal, one- to eight-segmented short cerci, and an
absence of wings except for the reproductive caste (Fig. 241). Wings, when

Figure 241. Reproductive termites, as illustrated below, possess two pairs of
equal sized wings. These wings are lost after the dispersal flight.

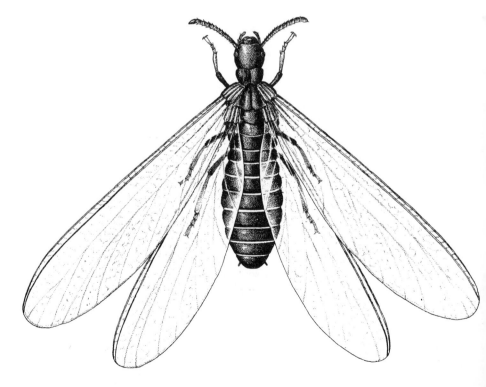

present, are longer than the body and are membranous. Fore and hindwings are similar in shape and size. All termites are social and their castes and metamorphosis are discussed in detail in the chapter on social insects.

Approximately 2,120 species are known (41 in North America) and are separated into 4 families. Often termed "white ants," this group is of great biological and economic importance. In the tropics and in forests their feeding recycles nutrients and aids in soil development. In other instances, however, their eating is in direct conflict with humans. Since Isoptera feed upon paper, wood, and other similar cellulose goods, they cause considerable damage. In many regions of the United States nearly every house is or has been infested in the past 10 years unless the soil was treated during building. Some of the more important pest species in the United States are the subterranean *Reticulitermes flavipes* (central and eastern states), *R. hesperus* (Pacific states), *R. tibialis* (west of Mississippi River), and the dry-wood *Incisitermes minor* (Pacific states), *I. hubbardi* (southwestern states), and *I. snyderi* (southeastern states).

DERMAPTERA

Earwigs (Fig. 242) are medium-sized insects (from 5 mm to 35 mm) with characteristic enlarged unsegmented forcep-like cerci. They have prognathic chewing mouthparts, filiform antennae, well-developed compound eyes, and long legs with three-segmented tarsi. The forewings are thickened into tegmina and are short. The semicircular hind wings fold fan-like longitudinally and twice transversely to fit under the reduced tegmina when not used for flight.

Eggs are often deposited in burrows where they are cleaned and protected by the female. Metamorphosis is incomplete. Nymphs feed, as do the adults, on a wide variety of materials from dead to living plants and animals. Activity is normally at night.

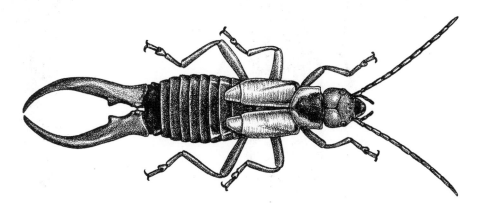

Figure 242.   An earwig (*Forficula auricularia*).

Approximately 1,000 species are known, but only 18 have been found in North America. Four families exist. Although a few species occasionally feed on cultivated plants, generally they are of little economic importance.

PLECOPTERA

Stoneflies (Fig. 243) are medium to large insects (from 12 mm to 65 mm in length) and are characterized by long filiform antennae, chewing mouthparts, cerci that are usually long and multisegmented, moderate to small compound eyes, and normally two pairs of wings with many veins. Forewings

Figure 243.   Stoneflies. (A) adult; (B) nymph.

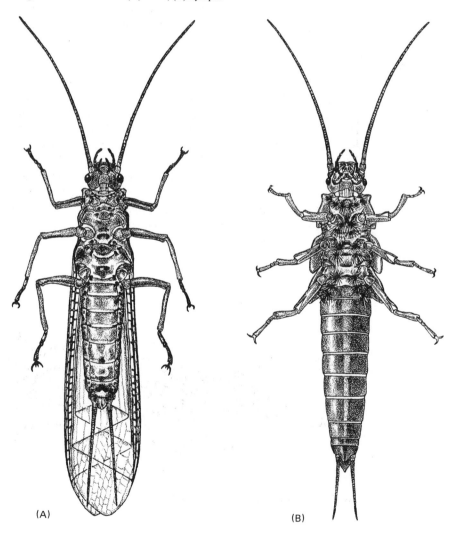

(A)

(B)

are narrower than the hindwings and both pairs fold flat over the abdomen when not in flight.

Metamorphosis is incomplete. Eggs are mostly oviposited in moving water. Nymphs (Fig. 243B) have long antennae, chewing mouthparts, and two long cerci. Gills, when present, are on the thorax, neck, or first two to three abdominal segments. As the name stonefly implies, nymphs are common on or under stones. Some species are herbivorous, but others are scavengers or carnivorous. Maturation takes from 1 to 4 years, at which time nymphs leave water and molt on nearby vegetation. Emergence from water occurs in the summer but some species normally accomplish this in the fall and winter.

The nearly 1,500 species (430 in North America) are classified into 6 to 10 families. They form an important food for many fish but are of no direct medical or economic importance to humans.

PSOCOPTERA

Psocids are small (from 1 mm to 7 mm) and have the following diagnostic characteristics: compound eyes large or reduced, chewing mouthparts, long filiform antennae, slender legs with two- or three-segmented tarsi, and no cerci. Wings may be lacking or present. When developed, the hindwings are smaller than the front pair; both pairs flex rooflike over the abdomen when not in use and have few veins.

Eggs are laid singly or in groups, either in buildings or outdoors. Nymphs feed, as do the adults, on paper, book bindings, glue, pollens, feathers, starch, grain, algae, or fungi. Some of the wingless species are called book lice (Fig. 244).

Figure 244. An adult book louse (*Liposcelis terricolus*).

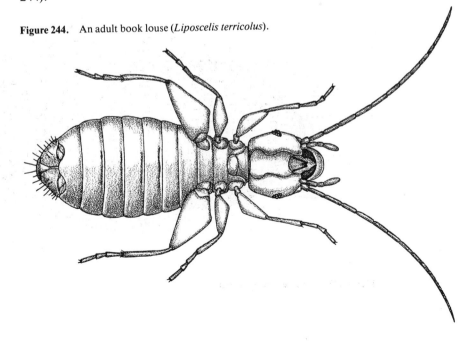

The 1,100 world species (150 in North America) are divided into 11 families. Some species are serious pests of stored cereal foods and libraries, the latter indicated by the name book lice. Most species, however, have no economic importance.

MALLOPHAGA

Chewing lice (Fig. 245) are small (from 2 mm to 6 mm) and have a head usually broader than long, modified chewing mouthparts, reduced compound eyes, two- to five-segmented tarsi, no cerci, and lack wings. The body is flattened dorsoventrally.

Eggs are fastened to feathers or hair of the host. Metamorphosis is incomplete. Both nymphs and adults ingest dead skin, feathers, hair, or scabs. Under high population pressures the dermal skin layer also may be attacked, particularly around wounds.

There are 2,675 species (318 in North America) and these are divided into 6 families. Most chewing lice infest birds although a few utilize mammals as a host. Host specificity is marked, and transfer from one host to another normally occurs only between two birds of the same species as the birds mate or nest. If a host dies, the louse fauna usually perishes. This order is of economic importance when domestic animals become infested; over 40 species are known to parasitize poultry and 7 species are very common. A few species infest livestock (horses, cattle, and goats). Loss of weights and lowered egg production, in the case of birds, are two common results of infestations.

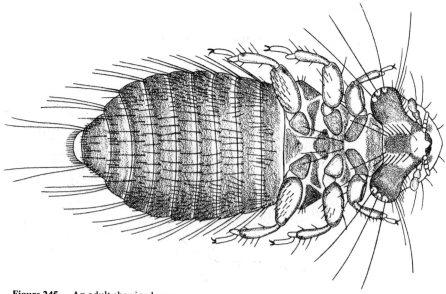

Figure 245.   An adult chewing louse.

ANOPLURA

Sucking lice (Fig. 246) are minute to small (from 0.4 mm to 6.5mm) and may be characterized by their narrower than long head, two- to five-segmented antennae, piercing–sucking mouthparts that are retracted into the head, greatly reduced eyes, absence of wings and cerci, and dorsoventrally flattened body. The legs are short and the single tarsus and claw are modified into a grasping organ by being apposed to a tibial lobe.

Sucking lice feed on blood and their entire life cycle is spent on mammalian hosts. Metamorphosis is incomplete. Eggs are glued to hair of the host. A high degree of host specificity and preference for specific regions on the host is recognizable. The human louse, *Pediculus humanus,* infests humans and whether it feeds in the head or body region has direct influence on its morphology and behavior (these two varieties were once considered two species). The crab louse, *Pthirus pubis,* is found mainly in the pubic and perianal region of humans.

Only 250 species are known (62 in North America) and these are in 3 families. One family includes lice found on humans, the second family infests only seals and walruses, and the third family includes all other species. Sucking

Figure 246.   An adult sucking louse. This species, *Pthirus pubis,* is normally found in the pubic region of man.

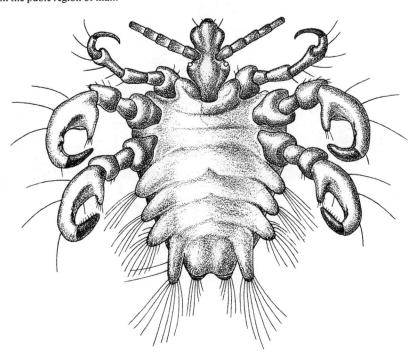

lice are important to humans in three ways: (1) several species infest livestock, (2) two species parasitize man and are irritating, and (3) human disease may be transmitted, including one type of relapsing fever, trench fever, and epidemic typhus. The latter two are normally associated with wars.

THYSANOPTERA

Thrips (Fig. 247) are minute to small (from 0.5 mm to 14 mm) and have long and narrow bodies, large compound eyes, six- to nine-segmented antennae, mouthparts modified for rasping–sucking in a cone-shaped beak, one- to two-segmented tarsi, and abdomen that is often pointed and lacking cerci. Wings are either absent or are long and narrow with long fringes of hair. Species may have a protrusible bladder at the apex of each tarsus. An appendicular ovipositor is present in some species.

Figure 247.   A female thrips (*Frankliniella tritici*).

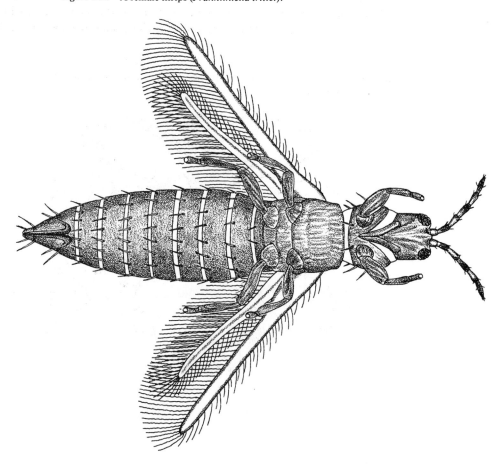

Metamorphosis is intermediate between incomplete and complete. The first two instars lack external wing pads and, although the instars resemble adults, they are often called larvae. The remaining instars have external wing pads, are inactive, and are defined as prepupa and pupae. Most active stages feed on plants although a few species are carnivorous. Parthenogenesis occurs in several species.

The 5,000 species (606 in North America) are separated into 5 families. They feed on a wide range of flowers and cultivated crops and cause considerable damage. Examples include tobacco, onion, gladiolus, and citrus. A few plant viruses are transmitted by thrips.

HEMIPTERA

The true bugs (Figs. 248, 249) vary in length up to 100 mm, compound eyes are usually large, antennae are from four- to five-segmented and often longer than the head, mouthparts are piercing–sucking with the segmented beak arising from the anterior portion of the head, tarsi are one- to three-segmented, cerci are absent, and the wings are normally present and positioned flat over the abdomen when at rest, separated by an enlarged *scutellum*. The front pair of wings is usually thickened at the base and membranous apically to form a *hemelytron*. The hindwings are membranous and slightly shorter than the hemelytra. Great variation in legs exists.

Metamorphosis is incomplete. Food varies from the common herbivorous to carnivorous (ambush and most assassin bugs) and parasitic (bedbugs). Many nymphs and adults are terrestrial, but a significant number are aquatic. Stink glands are common.

A number of true bugs are of economic and medical importance. Pests of cultivated plants include the squash bug, *Anasa tristis* (attacks squash and pumpkins), tarnished plant bug, *Lygus lineolaris* (fruit and flowers), cotton fleahopper, *Psallus seriatus* (cotton), chinch bug, *Blissus leucopterus* (wheat, oats, and other grains), harlequin bug, *Murgantia histrionica* (cabbage and turnips), and the southern green stinkbug, *Nezara viridula* (wide variety of crops). Some species of assassin bugs are naturally infected with chagus disease, most of which belong to the genus *Triatoma* (Fig. 250). The infection may be transmitted to humans by rubbing the protozoan organism in *Triatoma* feces through the skin by scratching. Because of modern insecticides, bedbugs have nearly been eliminated as a pest of humans. A few species of Hemiptera, such as the giant water bugs and backswimmer, bite severely, and should be avoided.

This large order contains 44 families with 25,000 species (4,500 in North America). The key on page 257 will expedite identification of the more common families.

Figure 248. Examples of common Hemiptera. (A) water scorpion (Nepidae); (B) water boatman (Corixidae); (C) backswimmer (Notonectidae); (D) assassin bug (Reduviidae); (E) water strider (Gerridae); (F) plant bug (Miridae); (G) toad bug (Gelastocoridae); (H) ambush bug (Phymatidae); (I) coreid bug (Coreidae); (J) giant water bug (Belostomatidae); (K) lygaeid bug (Lygaeidae); (L) stinkbug (Pentatomidae).

Figure 249.   An assassin bug (*Arilus cristatus*). This species is referred to as a wheel bug because of notched tergal region. It is a predator on many insects and has a painful bite.

Figure 250.   *Triatoma sanguisuga,* a reduviid, is a bloodsucking bug that feeds on mammals including, occasionally, humans.

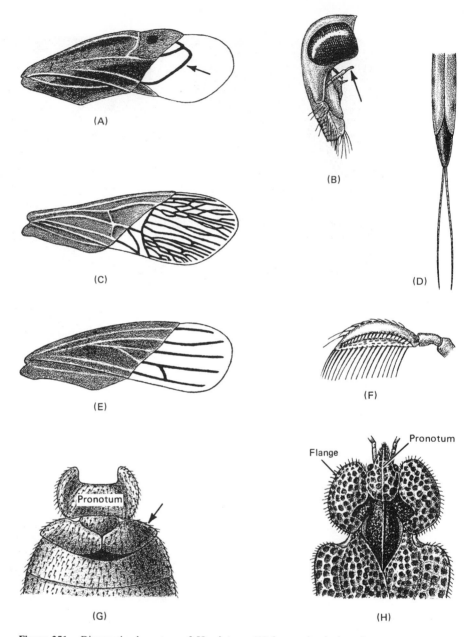

Figure 251. Diagnostic characters of Hemiptera. (A) loop veins in hemelytron (Miridae); (B) antenna reduced (Corixidae); (C) many veins in membrane of hemelytron (Coreidae); (D) long caudal appendages (Nepidae); (E) few veins in membrane of hemelytron (Lygaeidae); (F) foreleg spoonlike (Corixidae); (G) wings scalelike (Cimicidae); (H) thorax with lacelike pattern (Tingidae).

256

Key to Common Families of Hemiptera

1. Hind leg without tarsal claws 2
    Hind leg with tarsal claws 3

2. Front pair of legs spoon-shaped (Fig. 251F) . . . . . (water boatman) CORIXIDAE
    Front pair of legs not spoon-shaped, with claws
    . . . . (backswimmer) NOTONECTIDAE

3. Antennae easily seen, not reduced 6
    Antennae reduced and difficult to locate (Fig. 251B) 4

4. Abdomen with long slender caudal appendage (Fig. 251D)
    . . . . (water scorpion) NEPIDAE
    Abdomen lacking long slender caudal appendage 5

5. Body about as wide along (Fig. 248G); rough exoskeleton; 2 ocelli
    . . . . (toad bugs) GELASTOCORIDAE
    Body usually much longer than wide (Fig. 248J); no ocelli
    . . . . (giant water bugs) BELOSTOMATIDAE

6. Long narrow legs, hind femur extending much beyond tip of abdomen
    (Fig. 248E) . . . . (water striders) GERRIDAE
    Legs not unusually long, hind femur shorter than tip of abdomen or not much
    longer 7

7. Wingless or front wings reduced to short scales (Fig. 251G)
    . . . . (bedbugs) CIMICIDAE
    Wings fully developed 8

8. Pronotum with broad flanges (Fig. 251H); lacelike pattern in wings and pro-
    notum . . . . (lace bugs) TINGIDAE
    Not as above. 9

9. Loop veins in membrane of hemelytra (Fig. 251A) 10
    No loop veins in membrane of hemelytra (Fig. 251C) 12

10. Femur of front legs very thick, triangular-shaped (Fig. 252A)
     . . . . (ambush bugs) PHYMATIDAE
     Femur of front legs not thick or triangular 11

11. Beak curved, 3-segmented (Fig. 252B); front wings not tilted downward
     . . . . (assassin bugs) REDUVIIDAE
     Beak essentially sraight, 4-segmented; front wings tilted at distinct angle
     posterior to abdomen (Fig. 252C). . . . . (plant bugs) MIRIDAE

12. Scutellum greatly enlarged (Fig. 252D); antennae 5-segmented
     . . . . (stinkbugs) PENTATOMIDAE
     Scutellum not enlarged for order 13

13. Front wing membrane with 4-6 veins (Fig. 251E) . . . (lygaeid bugs) LYGAEIDAE
     Front wing membrane with more than 10 veins (Fig. 251C)
     . . . . (coreid bugs) COREIDAE

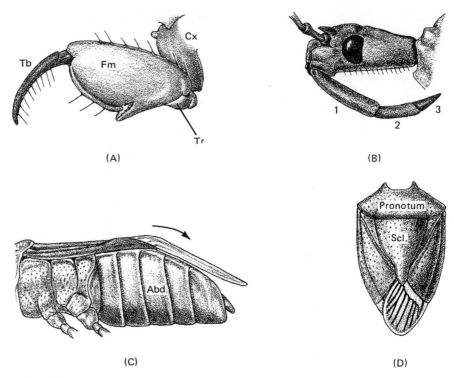

Figure 252. Diagnostic characters of Hemiptera. (A) thick femur (Phymatidae); (B) beak 3-segmented (Reduviidae); (C) wings tilted downward (Miridae); (D) scutellum greatly enlarged (Pentatomidae).

HOMOPTERA

This order (Fig. 253) includes cicadas, (Fig. 254), leafhoppers, treehoppers, spittlebugs, aphids, psyllids, and scales. Size ranges from minute to large (from 0.3 mm to 80 mm) with United States species, except the cicadas, measuring less than 12 mm. Diagnostic characters include large compound eyes, piercing–sucking mouthparts with the beak arising near the prothoracic legs, setaceous or filiform antennae, two- or three-segmented tarsi, wings absent or two pairs and membranous, and an absence of cerci. Females often possess an appendicular ovipositor (Fig. 52B).

These insects undergo incomplete metamorphosis, but life cycles may be very complex involving winged and wingless stages. Many aphids reproduce parthenogenetically and viviparous during the summer but are oviparous when fall arrives. Scales undergo a metamorphosis intermediate between incomplete and complete. Completion of life cycles varies from less than a week in some

Figure 253. Examples of Homoptera. (A) cicada (Cicadidae); (B) spittlebug (Cercopidae); (C) treehopper (Membracidae); (D) leafhopper (Cicadellidae); (E) planthopper (Fulgoridae).

Figure 254. An adult cicada (*Tibicen pruinosa*).

aphids to 17 years in a few cicadas (Figs. 255, 256). With the exception of cicadas, development takes place above ground.

All Homoptera are herbivorous and include many pests. Species of leafhoppers, white flies, scales, and aphids are especially detrimental to a wide variety of crops. Leafhoppers and aphids (Fig. 257) transmit over two-thirds of the known viral plant diseases of economic importance. Some species have become symbiotic with ants.

Figure 255. A cicada nymph (*Tibicen pruinosa*) with its large digging front pair of legs.

Figure 256. The emergence of an adult cicada (*Tibicen pruinosa*) from the exoskeleton of the last instar nymph.

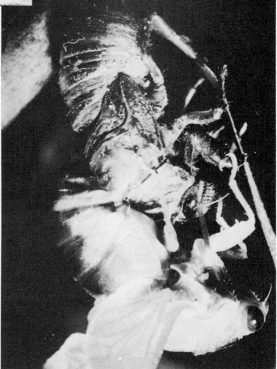

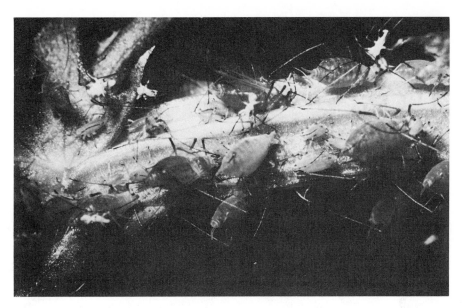

Figure 257. An aphid (*Macrosiphum pisi*). Both winged and wingless individuals are present.

Over 33,000 (6,700 in North America) species are known and grouped into about 32 families (varies with classification). The most common families may be determined by using the following key.

Key to the Common Families of Homoptera

1. Setaceous antennae (Fig. 258A) 2
   Filiform antennae. 6
2. Over ½ in. (1.27 cm); 3 ocelli . . . . (cicadas) CICADIDAE
   Less than ½ in. (1.27 cm); 2 ocelli or more 3
3. Pronotum enlarged dorsally (Fig. 258C) . . . . . . (treehoppers) MEMBRACIDAE
   Pronotum not enlarged dorsally 4
4. Antennae arise beneath or behind eyes (Fig. 258D)
   . . . . (planthoppers) FULGORIDAE
   Antennae arise between eyes (Fig. 258C). 5
5. Hind tibia with 1 or more rows of numerous spines (Fig. 258B)
   . . . . (leafhoppers) CICADELLIDAE
   Hind tibia without such rows of numerous spines
   . . . . (spittlebugs) CERCOPIDAE
6. Abdomen usually with cornicles (Fig. 258E); body rounded
   . . . . (aphids) APHIDIDAE
   Abdomen lacking cornicles; body not rounded 7
7. Bodies covered by waxy scale . . . . (scales) COCCIDAE
   Bodies not covered by waxy scale . . . . (jumping plant lice) PSYLLIDAE

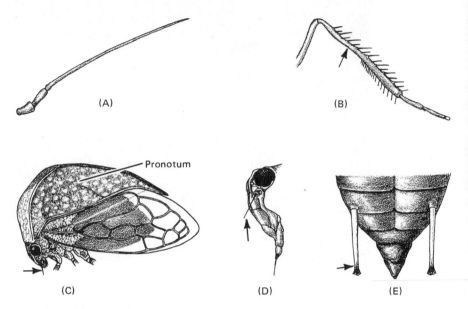

Figure 258.   Diagnostic characters of Homoptera. (A) setaceous antenna (Cicadellidae); (B) hind tibia with row of spines (Cicadellidae); (C) pronotum enlarged dorsally (Membracidae); (D) antenna below eye (Fulgoridae); (E) cornicles (Aphididae).

NEUROPTERA

These insects (Figs. 259A, 259B, 259C) vary in length—from 3 to 65 mm—and are characterized by having compound eyes large and widely separated, antennae usually filiform, chewing mouthparts, long legs with five-segmented tarsi, lacking cerci, and having two pairs of many veined membranous wings that are similar in size and usually held rooflike over the body when at rest (Fig. 260).

Metamorphosis is complete. Larvae are either aquatic (dobsonflies, alderflies) or terrestrial (mantispids, lacewings, ant lions). Aquatic forms (Fig. 261) have abdominal gills, strong chewing mouthparts, and are carnivorous. Pupation occurs outside the water in an earthen cell (Fig. 109) and the pupa is exarate. In contrast, terrestrial species (Fig. 262B) often have the mandible and maxilla from each side fuse to form a pair of hollow tubes for sucking fluids from prey. Cone-shaped holes in loose sand may be constructed to trap insects (Fig. 262A). The rarely seen mantispid larvae are parasitic on ground spider eggs.

Approximately 4,670 species are known (338 in North America) and these are placed into 14 or 15 families. A few species are beneficial to humans, for example, the green lacewings that feed on such pests as aphids and other Homoptera. None appears to be detrimental.

Figure 259. Examples of some minor insect orders. (A) Neuroptera (dobsonfly); (B) Neuroptera (green lacewing); (C) Neuroptera (ant lion); (D) Trichoptera (caddisfly); (E) Mecoptera (scorpionfly).

Figure 260. An adult green lacewing (*Chrysopa carnea*).

Figure 261. A larval dobsonfly (*Corydalus cornutus*), often termed hellgramite. Its aquatic existence may be deduced by the abdominal gills. Food consists of mayfly and stonefly nymphs and caddisfly larvae.

Figure 262. Ant lion larvae are found in sandy soil. (A) the coneshaped depressions constructed to facilitate prey slipping to the bottom where the larva waits. (B) the larva with its sickle-shaped mouthparts used in capturing and sucking fluids from the prey.

264

COLEOPTERA

Beetles (Figs. 263, 264) are short to long (from 0.25 mm to 150 mm) and include the largest insects by biomass known. Beetles are characterized by having chewing mouthparts, large compound eyes, lacking cerci, and usually possessing thickened forewings or *elytra,* which meet at the midline when folded. The membranous hindwings are the flight organs and fold up under

Figure 263. Examples of Coleoptera. (A) tiger beetle (Cicindelidae); (B) firefly (Lampyridae); (C) leaf beetles (Chrysomelidae); (D) ground beetle (Carabidae); (E) scarab beetle (Scarabeidae); (F) long-horned beetle (Cerambycidae); (G) soldier beetle (Cantharidae); (H) rove beetle (Staphylinidae); (I) click beetle (Elateridae); (J) weevil (Curculionidae); (K) ladybird beetle (Coccinellidae).

Figure 264. Examples of Coleoptera. (A) carrion beetle (Silphidae); (B) metallic wood borer (Buprestidae); (C) passalid beetle (Passalidae); (D) blister beetle (Meloidae); (E) stag beetle (Lucanidae); (F) darkling beetle (Tenebrionidae); (G) predaceous diving beetle (Dytiscidae); (H) water scavenger beetle (Hydrophilidae); (I) whirligig beetle (Gyrinidae); (J) dermestid beetle (Dermestidae).

the elytra when not in use. Antennae are highly variable. Tarsi are primitively five-segmented, but the number of segments may be less and are used in identification. Bodies are normally stout and sclerotization is extensive (Figs. 265, 266).

Metamorphosis is complete. Larvae are highly diverse in structure (Fig. 267) and in diets. Some are heavily sclerotized (click beetles, ground beetles) while others have poor sclerotization (scarabs and weevils). Larvae of many beetles such as predaceous diving beetles, whirligig beetles, and water scavenger beetles inhabit water, but most are terrestrial. A few are legless (borers). Pupae are exarate. Several years may be required for development.

Figure 265.   Cottonwood borer adult (*Plectrodera scalator*).

Figure 266.   Ground beetle adult (*Harpalus caliginosus*).

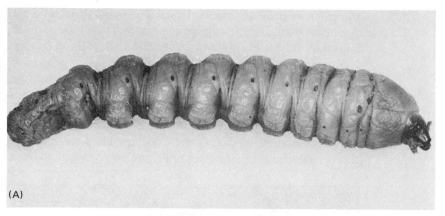

Figure 267. Examples of beetle larvae. (A) Cerambycidae; (B) Tenebrionidae; (C) Scarabeidae.

The larval stage is often the most destructive but many adults are also of economic importance. Stored products are attacked by the rice weevil *(Sitophilus oryzae),* yellow mealworm *(Tenebrio molitor),* confused flour beetle *(Tribolium confusum),* red flour beetle *(Tribolium castaneum),* carpet beetle *(Anthrenus scrophulariae),* and the larder beetle *(Dermestes lardarius).* The boll weevil *(Anthonomus grandis)* is a great pest on cotton and the many rootworm species *(Diabrotica* sp.) can cause much damage to corn. Vegetables may be greatly affected by the striped cucumber beetle *(Acalymma vittatum),* Colorado potato beetle *(Leptinotarsa decemlineata),* Mexican bean beetle *(Epilachna varivestis),* and the bean weevil *(Acanthoscelides obtectus).* The Japanese beetle *(Popillia japonica)* is one of the more serious pests of plants in eastern United States. The larvae feed on roots of grasses and the adults on leaves of a variety of plants. It should be noted, however, that a few species, including the ladybird beetles, are highly beneficial as predators of other detrimental insects such as aphids.

The order Coleoptera includes approximately 40 percent of all species of insects. An estimated 290,000 world and 28,600 species in North America exist. The number of families varies but Borer and DeLong (1976) cite 125. The following key is useful in identifying the most common adults of these families.

Key to the Common Families of Coleoptera

1. Each eye divided into dorsal and ventral halves (Fig. 268D); 2nd and 3rd pair of legs short . . . . (whirligig beetles) GYRINIDAE
   Eyes not divided; legs not short 2
2. Legs modified for swimming (Fig. 50D) 3
   Legs not modified for swimming 4
3. Sternal keel present (Fig. 268E). . . (water scavenger beetles) HYDROPHILIDAE
   Sternal keel absent . . . . (predaceous diving beetles) DYTISCIDAE
4. Head usually prolonged into a snout (Fig. 268I); antennae elbowed
   . . . . (weevils) CURCULIONIDAE
   Head not prolonged into a snout; antennae usually not elbowed 5
5. Tarsal formula appears to be 5-5-5 10
   Tarsal formula either less or appearing to be less than 5-5-5 (one segment may be greatly reduced, Fig. 268H) 6
6. Tarsal formula 5-5-4 7
   Tarsal formula not 5-5-4 8
7. Each tarsal claw split longitudinally and appearing to be double
   (Fig. 268F) . . . . (blister beetles) MELOIDAE
   Tarsal claws not split longitudinally (Fig. 268G)
   . . . . (darkling beetles) TENEBRIONIDAE

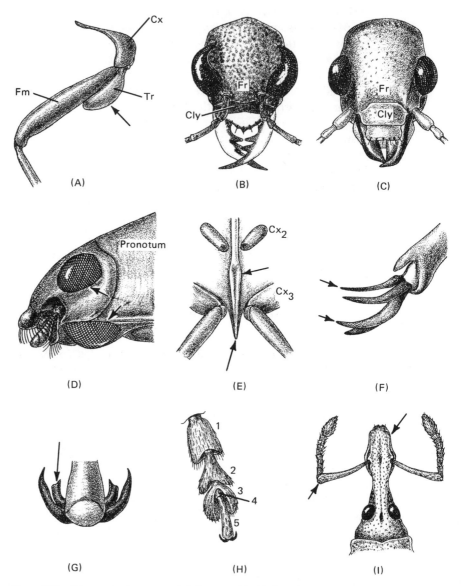

Figure 268. Diagnostic characters of Coleoptera. (A) hind trochanter with greatly enlarged lobes (Cicindelidae); (B) clypeus wide (Cicindelidae); (C) clypeus narrow (Carabidae); (D) eye divided into two halves (Gyrinidae); (E) sternal keel (Hydrophilidae); (F) split claws (Meloidae); (G) toothed claw (Cerambycidae); (H) tarsal formula appears to be four but is actually five (Chrysomelidae); (I) head prolonged, antenna elbowed (Curculionidae).

8. Tarsal formula appears to be 3–3–3 (actually 4–4–4)
    . . . . (ladybird beetles) COCCINELLIDAE
    Tarsal formula appears to be 4–4–4 (actually 5–5–5). 9
9. Eyes notched (Fig. 269A); antennae long, often longer than body
    (Fig. 265) . . . . (long-horned beetles) CERAMBYCIDAE
    Eyes not notched; antennae short . . . . (leaf beetles) CHRYSOMELIDAE

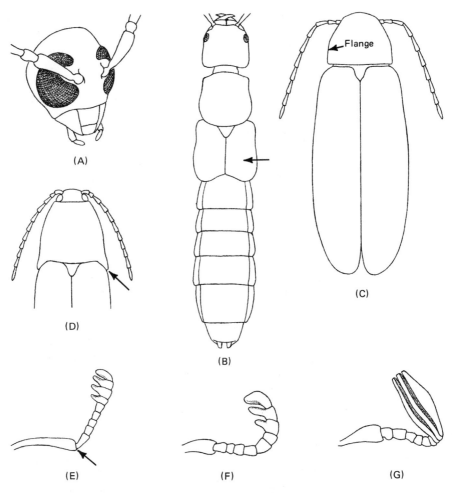

Figure 269. Diagnostic characters of Coleoptera. (A) notched eye (Cerambycidae); (B) elytra short and with five or more terga exposed (Staphylinidae); (C) pronotum enlarged and covering head (Lampyridae); (D) pronotum with sharp posterior angles to flange (Elateridae); (E) antenna elbowed but poorly lamellated (Lucanidae); (F) antenna not elbowed, poorly lamellated (Passalidae); (G) antenna not elbowed not lamellated (Scarabeidae).

10. Hind trochanters with greatly enlarged lobes (Fig. 268A) 11
    Hind trochanters without such enlarged lobes. 12

11. Clypeus wider than distance between antennal sockets (Fig. 268B); head usually broader than prothorax . . . . (tiger beetles) CICINDELIDAE
    Clypeus narrower than distance between antennal sockets (Fig. 268C); head usually narrower than prothorax . . . . (ground beetles) CARABIDAE

12. Elytra short, 3 or more terga exposed (Fig. 269B) 13
    Elytra covering the abdomen 14

13. Three terga exposed . . . . (carrion beetles) SILPHIDAE
    Five or more terga exposed (Fig. 269B) . . . . . . (rove beetles) STAPHYLINIDAE

14. Pronotum enlarged and with lateral flanges; head difficult to see from dorsal view (Fig. 269C). . . . . (fireflies) LAMPYRIDAE
    Pronotum without such enlarged flanges 15

15. Pronotum with sharp posterior angles at lateral corners (Fig. 269C)
    . . . . (click beetles) ELATERIDAE
    Pronotum without sharp lateral angles 16

16. Antennae elbowed (Fig. 269E); mandibles greatly enlarged in males
    . . . . (stag beetles) LUCANIDAE
    Antennae not elbowed (Figs. 269F, 269G). 17

17. Antennae highly lamellate (Fig. 269G) . . . . . . (scarab beetles) SCARABAEIDAE
    Antennae not lamellate or only poorly so (Fig. 269F) 18

18. Large black shiny beetles; often "horn" on head
    . . . . (passalid beetles) PASSALIDAE
    Not as above. 19

19. Rounded body with long legs; legs with yellow hair . . (spider beetles) PTINIDAE
    Not as above. 20

20. Ventral surface with dense white hair . . . . . (dermestid beetles) DERMESTIDAE
    Little if any ventral hair 21

21. Elytra poorly sclerotized and flexible; usually elytra yellow or orange with black markings (Fig. 263G) . . . . (soldier beetles) CANTHARIDAE
    Elytra strongly sclerotized and hard . . . . (metallic wood borers) BUPRESTIDAE

HYMENOPTERA

This order includes bees, wasps, ants, and sawflies (Figs. 270, 271). Sizes range from 0.21 mm to 65 mm in length, excluding the appendicular ovipositor (Fig. 272). Characteristics include filiform antennae, chewing or chewing-lapping mouthparts, large compound eyes except for ants, long legs with five-segmented tarsi, cerci minute or absent, and wings absent (Fig. 273) or two pair which are long and narrow with fused venation. The ovipositor may be modified into a sting (Fig. 53). Some species are social and have caste differences (see chapter on social insects).

Figure 270. Examples of Hymenoptera. (A) sawfly (Tenthredinidae); (B) braconid wasp (Braconidae); (C) ichneumonid wasp (Ichneumonidae); (D) horntail (Siricidae); (E) cimbicid sawfly (Cimbicidae); (F) ants (Formicidae).

Figure 271. Examples of Hymenoptera. (A) vespid wasp (Vespidae); (B) velvet ant (Mutillidae); (C) honeybee (Apidae); (D) sweat bee (Halictidae); (E) leafcutting bee (Megachilidae); (F) spider wasp (Pompilidae); (G) bumblebee (Apidae); (H) sphecid wasp (Sphecidae); (I) carpenter bee (Xylocopidae).

Figure 272. A horntail (*Tremex columba*).

Figure 273. A female ant (*Dasymutilla occidentalis*). Larvae feed on other wasp larvae.

Metamorphosis is complete. Larvae are either caterpillar-like as in the sawflies (Fig. 274) or vermiform in the remainder (Fig. 275). Caterpillars are herbivorous, but other larvae vary in diet from parasitic to scavengers. Adults feed daily or mass provision nests with sufficient food for the larvae to complete development. Pupae are exarate and sometimes enclosed within a cocoon.

Relatively few species are considered pests. Wasps (Fig. 179), bees (Fig. 187), and some ants such as the southern fire ant *(Solenopsis xyloni)* are well

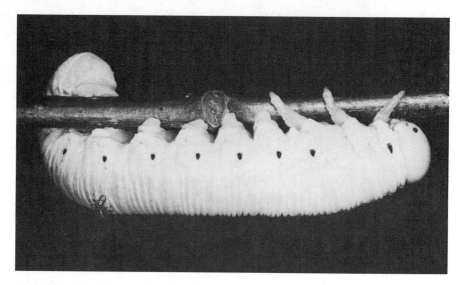

Figure 274.   The elm sawfly (*Cimbex americana*) larva.

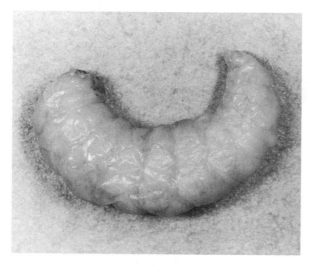

Figure 275.   A bumblebee larva (*Bombus americanorum*).

known for their stings. Of conspicuous economic importance to crops are the pearslug *(Caliroa cerasi),* larch sawfly *(Pristiphora erichsonii),* wheat stem sawfly *(Cephus cinctus),* wheat jointworm *(Harmolita tritici),* and the Texas leaf-cutter ant *(Atta texana).* Beneficial species should be expecially noted, however. Some of the major pollinators such as the honeybee *(Apis mellifera)* and biological control agents of other insects are Hymenoptera. Ichneumonid, braconid, and chalcid wasps, for example, deposit their eggs on many aphids and lepidopteran caterpillars that are pests of humans.

The 113,000 species (17,300 in North America) are divided into 71 families. The following key to adults permits identification of the more common or conspicuous families in the United States.

Key to the Common Families of Hymenoptera

1. Slender waist (petiole) between abdominal segments 1 and 2 absent
   (Fig. 276A) . . . . suborder SYMPHYTA, 2
   Slender waist (petiole) between abdominal segments 1 and 2 present
   (Fig. 276B) . . . . suborder APOCRITA, 4

Figure 276. Diagnostic characters of Hymnoptera. (A) abdominal segments not narrowed (Tenthredinidae); (B) abdominal segments narrowed to form a waist or petiole (Sphecidae); (C) two recurrent veins present in forewing (Ichneumonidae); (D) one recurrent vein in forewing (Braconidae).

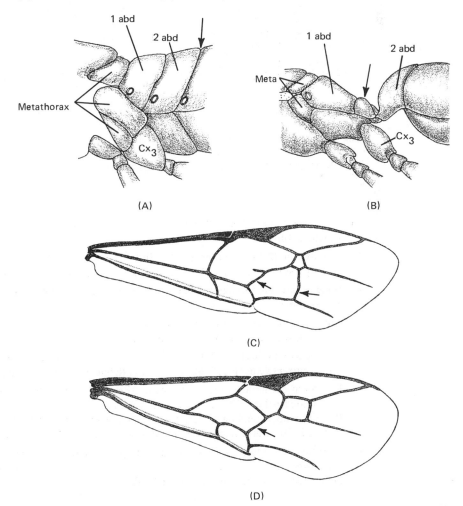

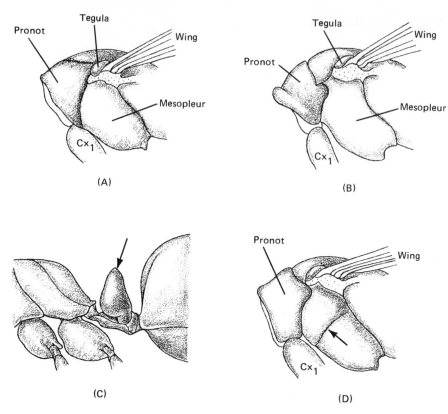

Figure 277. Diagnostic characters of Hymenoptera. (A) pronotum extending to tegula (Vespidae); (B) pronotum not extending to tegula (Sphecidae); (C) petiole with node or spine (Formicidae); (D) mesopleuron with transverse suture (Pompilidae).

2. Antennae clubbed (Fig. 270E) . . . . (cimbicid sawflies) CIMBICIDAE
   Antennae filiform 3

3. Usually 1 in. (2.54 cm) or more in length . . . . (horntails) SIRICIDAE
   Usually less than ½ in. (1.27 cm) in length . . . . . (sawflies) TENTHREDINIDAE

4. Wing venation reduced to a single anterior vein . . (chalcid wasps) CHALCIDAE
   Not as above. 5

5. Antennae long, with more than 16 segments 6
   Antennae with fewer than 16 segments 7

6. One recurrent vein (Fig. 276D) . . . . (braconids) BRACONIDAE
   Two recurrent veins (Fig. 276C) . . . . (ichneumonids) ICHNEUMONIDAE

7. Pronotum extending posterior to tegula (Fig. 277A); may be wingless or hairy
   and slender (Fig. 273) 8
   Pronotum not extending posterior to tegula, posterior edge of pronotum lobe-like
   (Fig. 277B); winged; may be hairy and stout bodied 11

8. Segment in petiole with a dorsal bump or spine (Fig. 277C); antennae
   elbowed . . . . (ants) FORMICIDAE
   Petiole without bumps or spines (Fig. 276B) 9

9. Usually covered by numerous hair (Fig. 273) . . . . (velvet ants) MUTILLIDAE
    Usually little conspicuous hair 10
10. Mesopleuron divided with straight, transverse suture (Fig. 277D); hind femora
    extend to near tip of abdomen (or longer) . . . . . (spider wasps) POMPILIDAE
    Not as above; wings folded longitudinally when flexed . . . . (vespids) VESPIDAE
11. With branched and plumose hair present . . . . (bees) 12
    Without branched and plumose hair present . . . . . (sphecid wasps) SPHECIDAE
12. Thorax usually metallic green . . . . (mining or sweat bees) HALICTIDAE
    Thorax not metallic green 13
13. Dense hair (yellow) only on 1st abdominal segment; rest of abdomen black or
    metallic blue (Fig. 271I). . . . . (carpenter bees) XYLOCOPIDAE
    Dense hair on more than 1st abdominal segment (Figs. 271C, 271G)
    . . . . (apid bees) APIDAE

MECOPTERA

Scorpionflies (Fig. 259E) are medium-sized (from 10 mm to 25 mm) and have the following important characteristics: chewing mouthparts are located at the end of a snout-like beak (Fig. 278), antennae are filiform and long, compound eyes are large and well-separated, legs are long and tarsi are five-segmented, cerci are simple in males and are two-segmented in females, and two pairs of membranous wings (lacking in some species) are of similar size and shape. Male genitalia are enlarged, bulblike, and held scorpion-like in repose above the abdomen.

Metamorphosis is complete. Eggs are normally deposited in the soil and the caterpillar-like larvae feed as predators or as scavengers on dead and

Figure 278.   A female scorpionfly (*Panorpa* sp.).

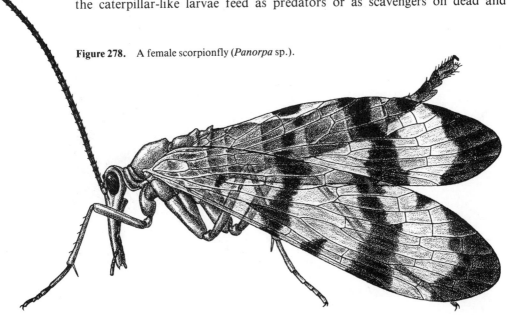

decaying materials. Pupation takes place in a cell in the soil. Adults are omnivorous and feed on such a variety of food as pollen, nectar, juices from decaying fruit, and dead or live insects. One group, the snow scorpionflies, are active throughout the year.

Only 410 species (80 in North America) are known and these are relegated to 4 families. Species are of no economic or medical importance to humans.

TRICHOPTERA

Caddisflies (Figs. 259D, 279) are small to moderate in size (from 1.5 mm to 40 mm). The compound eyes are large, mouthparts are chewing but poorly developed except for the four- to five-segmented maxillary palpi, antennae are long and filiform, legs are long and tarsi are five-segmented, cerci are one- or two-segmented, and two pairs of membranous wings are held roof-like over the abdomen. Hind wings are wider than the forewings, few cross veins are present in either set of wings, and hair is common on the wing surface.

Metamorphosis is complete. Eggs are deposited in gelatinous secretions in water or on objects at the water's edge. Larvae have chewing mouthparts, abdominal gills, and a pair of terminal hooks for aiding in locomotion and holding on to rocks or to a portable case (Fig. 280). These cases are distinctive with each species using characteristic materials and shape. Those that lack cases produce a web of silk that traps organic matter passing in the current. Pupation takes place in the case. The exarate pupa, equipped with sharp mandibles that will be lost after metamorphosis, cuts its way free of this case, swims to the shore, and molts into the adult.

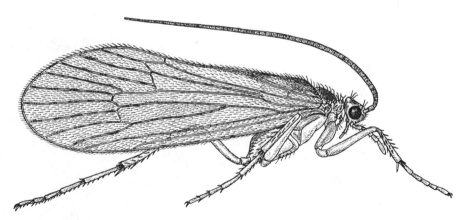

Figure 279.   A caddisfly adult.

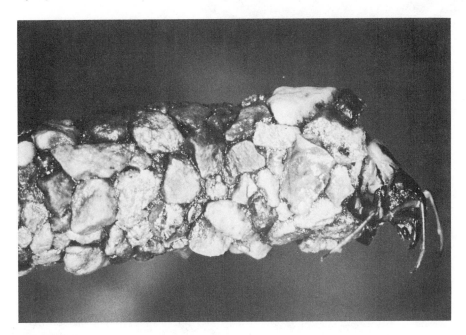

Figure 280.   A larval caddisfly with its pebble case.

Some 4,600 species are known (900 in North America) and these are divided into 20 families. Although they are part of many aquatic food chains, caddisflies cannot be considered to be of direct economic importance.

LEPIDOPTERA

Moths (Fig. 281) and butterflies (Fig. 282) vary in wingspread from 5 mm to 270 mm. This conspicuous order may be identified by siphoning mouthparts coiled under their head (Fig. 283), vestigial or absent maxillary palpi, large labial palpi, large compound eyes, long legs with five-segmented tarsi, absence of cerci, two pairs of membranous wings clothed with overlapping scales, and a body covered with hair and scales. Color is the result of not only pigments in the hair and scales but also from structural ridges and layers that reflect light differentially to cause irridescence. Antennae vary greatly and are useful in identification.

Metamorphosis is complete. Eggs are often deposited on specific host plants. Larvae, called caterpillars, have chewing mouthparts and up to five pairs of abdominal prolegs provided with hooks or *crochets*. Many larvae have numerous hairs and are brightly colored. A few have stinging or urticating hair. Cocoons may be spun by moth larvae, but most butterflies lack this

Figure 281. Examples of moths. (A) Sphingidae; (B) Pyralidae; (C) Yponomeutidae; (D) Notodontidae; (E) Arctiidae; (F) Noctuidae; (G) Lasiocampidae.

Figure 282. Examples of butterflies. (A) Nymphalidae; (B) Papilionidae; (C) Hesperidae; (D) Lycaenidae; (E) Danaidae; (F) Pieridae; (G) Libytheidae.

Figure 283.   A butterfly.

structure. Pupae are normally obtect. The dominant feeding stage is the larva (Figs. 284, 285); when the adults feed, they ingest nectar, but some utilize only stored fat from the larva as an energy source.

Many species of Lepidoptera are of economic or medical importance. Forests are frequently damaged by the gypsy moth *(Porthetria dispar),* eastern tent caterpillar *(Malacosoma americanum),* and the forest tent caterpillar *(Malacosoma disstria).* Caterpillars that damage fruit trees include the peach tree borer *(Sanninoidea existiosa),* peach twig borer *(Anarsia lineatella),* Oriental fruit moth *(Grapholitha molesta),* and the codling moth *(Laspeyresia pomonella).* The following are serious pests of grain: European corn borer *(Ostrinia nubilalis),* southwestern corn borer *(Diatraea grandiosella),* corn earworm *(Heliothis zea),* fall armyworm *(Spodoptera frugiperda),* armyworm *(Pseudaletia unipuncta),* several species of cutworms, and the sugarcane borer *(Diatraea saccharalis).* Cotton is damaged by the pink bollworm *(Pectinophora gossypiella);* tobacco is frequently eaten by the tobacco hornworm *(Manduca sexta).* The imported cabbage worm *(Pieris rapae)* is a pest on cabbage and related plants. Bagworms *(Thyridopteryx ephemeraeformis)* are

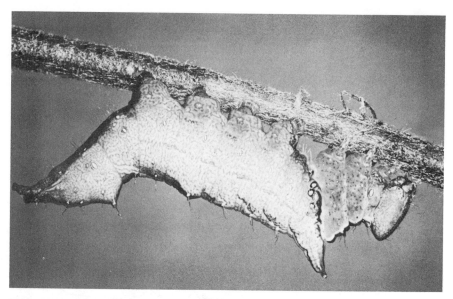

Figure 284. This lepidopteran caterpillar (*Schizura unicornis*) feeds on a wide variety of plants including apple, cherry, plum, willow, elm, oak, locust, and dogwood.

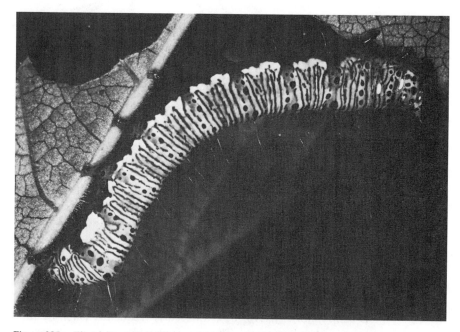

Figure 285. The eight-spotted forester moth larva (*Aloypia octomaculata*) has a narrow food preference and feeds mainly on grape and Virginia creeper.

common on evergreen. Stored products are often infested with Indian-meal moth *(Plodia interpunctella),* Angoumois grain moth *(Sitotroga cerealella),* Mediterranean flour moth *(Angasta kuhniella),* and two species of clothes moths *(Tineola bisselliella* and *Tinea pellionella).* Several species are urticating including the puss caterpillar *(Megalopyge opercularis)* and the Io moth caterpillar *(Automeris io).*

This large order includes 114,000 species (11,000 in North America), which are grouped into 77 families. Identification relies heavily on wing venation, and, since wings are covered with scales, viewing requires special bleaching techniques. To avoid this difficulty, a key has been prepared that includes only those distinctive families that are easily recognizable by color and body shape.

Key to the Common Families of Lepidoptera

1. Antennae clubbed (Fig. 283). 2
   Antennae not clubbed . . . . (moths) 6

2. Terminal hook usually projecting from clubbed antenna (Fig. 286A)
   . . . .(skippers) HESPERIIDAE
   Terminal hook lacking . . . . (butterflies) 3

3. Prothoracic legs much reduced (Fig. 286C)
   . . . . (brush-footed butterflies) NYMPHALIDAE
   Prothoracic legs normal size. 4

4. Large, wingspread usually over 2 in. (5.1 cm); hind wings with "tails" (Fig. 282B); prothoracic legs bearing tibial epiphyses (Fig. 286D)
   . . . . (swallowtails) PAPILIONIDAE
   Not as above. 5

5. Margin of eye notched where antenna located (Fig. 286E)
   . . . . (blues, coppers, etc.) LYCAENIDAE
   Margin of eye without notch . . . . (whites and sulphurs) PIERIDAE

6. Eye spots usually on wings; usually wingspread over 3 in. (7.6 cm)
   . . . . (giant silkworm moths) SATURNIIDAE
   Eye spots usually absent 7

7. Antennae pectinate but not obviously to tip (Fig. 286B)
   . . . .(royal moths) CITHERONIIDAE
   Antennae either pectinate to tip (Fig. 131) or not conspicuously plumose. 8

8. Hindwings reduced in width, fitting into indented posterior margin of front wings (Fig. 286F) . . . . (sphinx moths) SPHINGIDAE
   Hindwings normal; no indented posterior margin in front wings (Fig. 286H) . . . 9

9. Abdomen colored other than brown; front wings either white or brightly striped . . . . (tiger moths) ARCTIIDAE
   Abdomen light yellow-brown to dark brown. 10

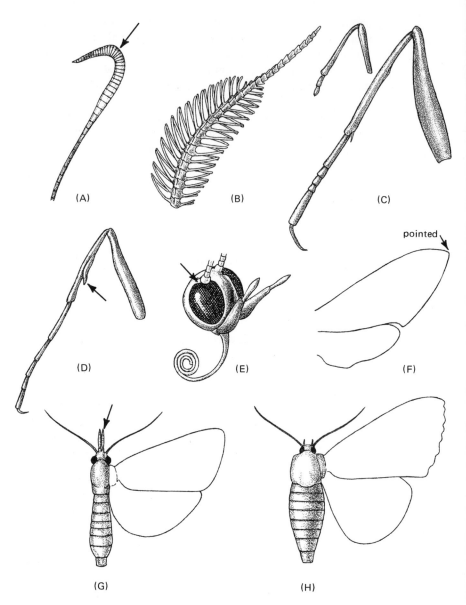

Figure 286. Diagnostic characters in Lepidoptera. (A) clubbed antenna with a hook (Hesperiidae); (B) antenna not plumose to tip (Citheroniidae); (C) prothoracic leg much smaller than mesothoracic (Nymphalidae); (D) tibial epiphysis (Papilionidae); (E) notched eye (Pieridae); (F) hindwing reduced in width, fitting into notch in pointed forewing (Sphingidae); (G) hindwing normal in width, body small, beaklike palps (Pyralidae); (H) hindwing normal in width, body large (Noctuidae).

10. Body large when compared to wing size; no large palpi extending beak-like in
     front of head. 11
     Body comparatively small; large beak-like palpi extending forward of head (Fig.
     286G) . . . . (pyralid moths) PYRALIDAE

11. Antennae distinctly pectinate
     . . . . (tent caterpillar moths) LASIOCAMPIDAE
     Antennae not distinctly pectinate . . . . (noctuid moths) NOCTUIDAE

SIPHONAPTERA

Fleas (Fig. 287) are minute to small (from 0.8 mm to 5 mm) and have the
following characteristics; compound eyes are absent or each is represented by a
single ommatidium, antennae are short and can be folded into grooves in the
head, mouthparts are piercing–sucking, coxae are long and tarsi are five-
segmented, cerci are small and one-segmented, and wings are absent. The body
is laterally flattened; posteriorly projecting rows of strong spines may be present, especially on the gena and pronotum where they may be enlarged to form
combs or *ctenidia*.

Metamorphosis is complete. Eggs are oviposited on the host or more
often in the host's nest; the former eggs fall off prior to hatching. The legless
larva (Fig. 288) feeds upon such organic matter as may be available including

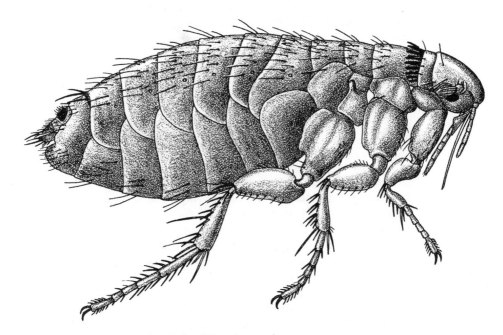

Figure 287.   An adult male flea (*Thrassis pansus*).

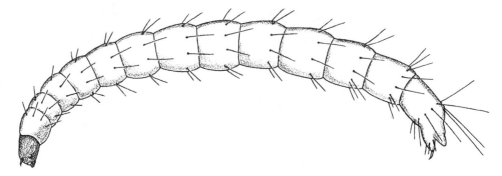

Figure 288.   A larval flea.

fecal material from adult fleas that contains blood residues. Pupation is in silken cocoons. Adults feed on blood from either birds or mammals, the latter being more common. Host specificity exists but is not as well developed as in lice. Some species are predominantly on the host but, if the host has a nest, many species of fleas leave during nonfeeding periods. These fleas, commonly called *nest fleas,* have their highest population in the nest, and scientific surveys only of hosts fail to reveal the true flea population.

The 1,600 species (248 in North America) are classified into 9 families. Beyond mere irritation, fleas are of medical importance to humans through disease transfer. Fleas are vectors of plague (bubonic form) and endemic or murine typhus. Several species of tapeworm can, but not commonly, infect humans after utilizing the flea as an intermediate host. In the tropics the chigoe flea attaches itself to humans and can initiate severe lesions. Fleas can also become pests to such domesticated animals as dogs and cats.

DIPTERA

True flies (Figs. 289, 290) are minute to medium-sized (from 0.5 mm to 50 mm) with wingspreads ranging from 1 mm to 100 mm. All are characterized by having only one pair of flight wings; the second pair is modified into halteres (Fig. 46). A great variety of mouthparts have evolved including piercing–sucking (Fig. 291), cutting–sponging, and sponging (Fig. 292). Labial palps are absent. Compound eyes are very large, antennae have many shapes, the mesothorax is disproportionately large, legs are long and tarsi are five-segmented, and cerci are absent.

Metamorphosis is complete. Larvae are normally legless and vermiform (Fig. 293), often without a well-developed head capsule (Fig. 226B). Great differences in diet exist and many habitats are exploited. Pupae may be active and nonfeeding (mosquitoes), but most pupae are sessile within puparia. A significant number of adults utilize blood or nectar for food.

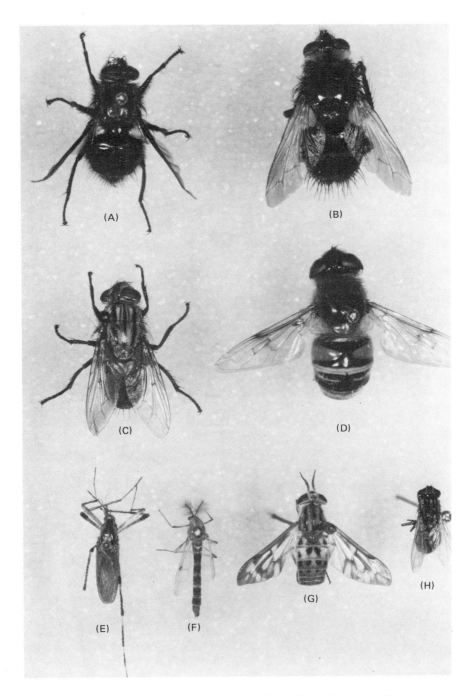

Figure 289. Examples of Diptera. (A) blowfly (Calliphoridae); (B) tachina fly (Tachinidae); (C) flesh fly (Sarcophagidae) (D) flower fly (Syrphidae); (E) mosquito (Culicidae); (F) midge (Chironomidae); (G) deer fly (Tabanidae); (H) house fly (Muscidae).

289

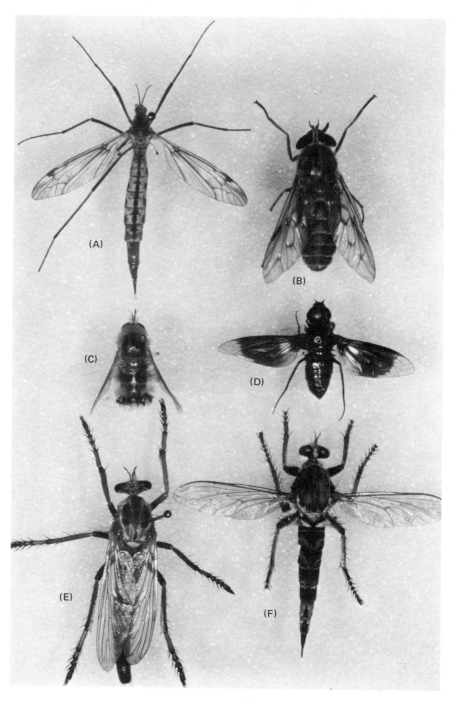

Figure 290. Examples of Diptera. (A) crane fly (Tipulidae); (B) horsefly (Tabanidae); (C) bee fly (Bombyliidae); (D) bee fly (Bombyliidae); (E) robber fly (Asilidae); (F) robber fly (Asilidae).

Figure 291. A robber fly (*Proctacanthella cacopiloga*).

Figure 292. A flesh fly (*Sarcophaga bullata*).

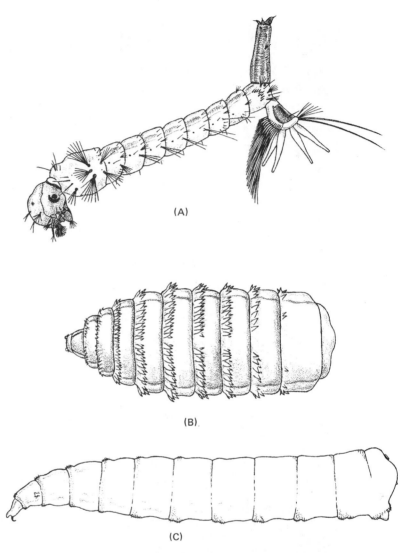

Figure 293. Examples of Diptera larvae. (A) mosquito (*Culex* sp.); (B) horse botfly (*Gasterophilus* sp.); (C) house fly (*Musca* sp.).

Numerous species are of economic and medical importance to humans. Injurious to plants as larvae are the Hessian fly (*Mayetiola destructor*) on wheat, sorghum midge *(Contarinia sorghicola)* on sorghum, and the Mediterranean fruit fly *(Ceratitis capitata)* on a wide number of fruits. Many larvae are parasitic on livestock including the screwworm *(Cochliomyia hominivorax)*, cattle grubs (*Hypoderma lineatum* and *H. bovis*), sheep bot fly

(Oestrus ovis), and common horse bot fly *(Gasterophilus intestinalis).* Adults may ingest blood from domesticated animals and include the horn fly *(Haematobia irritans),* stable fly *(Stomoxys calcitrans),* and many species of genera of horseflies, deer flies, black flies, and mosquitoes. Of prime importance to humans are the many diseases transmitted by mosquitoes, tsetse flies, and black flies, some of which are summarized in Table 3 (page 207). A few species, primarily in the families Syrphidae and Tachinidae, are of benefit to humans because of their parasitism of other insects.

Over 86,000 species (17,130 in North America) are described and separated into 105 families. The most conspicuous families may be identified by using the following key.

Key to the Common Families of Diptera

1. Antennae filiform to plumose, with at least 7 segments (Fig. 294A)
    . . . . suborder NEMATOCERA, 2
    Antennae with fewer than 7 segments (Figs. 294B, 294C) 4

2. Mesonotum with V-shaped suture above front edge of wings (Fig. 294D)
    . . . . (crane flies) TIPULIDAE
    Mesonotum without V-shaped suture 3

3. With long beak (Fig. 294E); scales on wing veins . . . . (mosquitoes) CULICIDAE
    Beak and scales on wings absent . . . . (midges) CHIRONOMIDAE

4. Arista absent (Fig. 294B). suborder BRACHYCERA, 5
    Arista present (Fig. 294C) . . . . suborder CYCLORRHAPHA, 7

5. Vertex indented between eyes (Fig. 294F) . . . . (robber flies) ASILIDAE
    Vertex not indented 6

6. Body with dense hair; head not broad when viewed above (Fig. 290C)
    . . . . (bee flies) BOMBYLIIDAE
    Body without dense hair; head broad when viewed above
    (Fig. 290B) . . . . (horse and deer flies) TABANIDAE

7. Spurious vein present in wing (Fig. 294G) . . . . (flower flies) SYRPHIDAE
    Spurious vein absent in wing. 8

8. Ridge or distinct bulge beneath and behind scutellum (Fig. 295A); arista
    bare . . . . (tachinid flies) TACHINIDAE
    No distinct ridge or bulge (Fig. 295B); arista at least partly plumose
    (Fig. 294C) 9

9. Hypopleuron without row of bristles . . . . (muscid flies) MUSCIDAE
    Hypopleuron with distinct row of bristles (Fig. 295C). 10

10. Notopleuron with 2 large bristles (Fig. 295D); arista plumose to tip
    (Fig. 294C) . . . . (blow flies) CALLIPHORIDAE
    Notopleuron with 4 large bristles (Fig. 295E); arista plumose only in basal
    half . . . . (flesh flies) SARCOPHAGIDAE

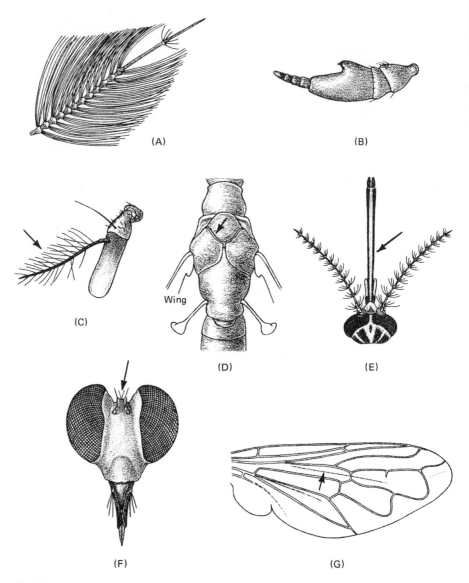

Figure 294. Diagnostic characters of Diptera. (A) antenna with more than seven segments and plumose (Culicidae); (B) antenna with less than seven segments, arista absent (Tabanidae); (C) antenna with less than seven segments, artista present and plumose to tip (Calliphoridae); (D) mesonotum with V-shaped suture (Tipulidae); (E) long proboscis (Culicidae); (F) vertex indented (Asilidae); (G) spurious vein present in wing (Syrphidae).

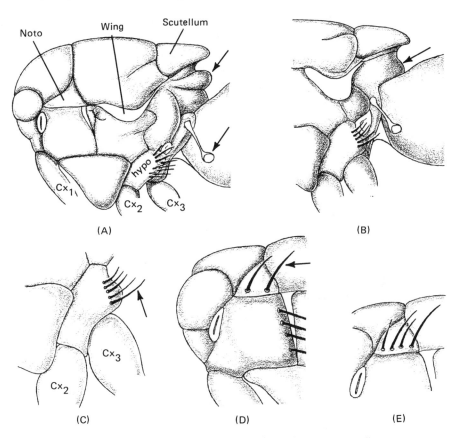

Figure 295. Diagnostic characters of Diptera. (A) distinct ridge behind scutellum (Tachinidae); (B) distinct ridge behind scutellum absent (Calliphoridae); (C) hypopleuron with distinct row of bristles (Calliphoridae); (D) notopleuron with two large bristles (Calliphoridae); (E) notopleuron with four large bristles (Sarcophagidae).

MAKING A REPRESENTATIVE COLLECTION OF INSECTS

Where and How to Collect

To secure a representative collection of insects during any season of the year, one must be constantly alert for different species. A collection built from specimens picked up at random or over a very short period is never representative of the species that occur in a locality. Therefore, make definite collecting trips to various habitats and localities to secure the best material. Large numbers of insects may be caught by turning over stones, logs, rubbish, and leaves. Sweepings in alfalfa, grass, and weed patches will reveal entirely different species. Butterflies may be caught by dropping a net (Fig. 296) over them when they alight. Water insects may be seined by using a water net. Various insects are attracted to sap exuding from stumps or tree trunks and furnish good collecting in the very early spring. Soil insects may be located by placing humus and leaf matter in a Berlese funnel or by using different sized sieves to sort material by size.

The list of habitats given below is by no means an exhaustive one, but it does include sites where representatives of orders may be found.

Apterygota (including springtails and bristletails)   Under old paper, leaves, old books, and in flour mills. Especially common in damp places such as under boards or decaying vegetation.

Ephemeroptera (mayflies), *Odonata* (damselflies, dragonflies), *Plecoptera* (stoneflies), and *Trichoptera* (caddisflies)   Adults near streams and ponds resting on rocks, reeds, bushes, or near lights. Immatures in streams, ponds, and other water habitats.

Orthoptera (grasshoppers, cockroaches, etc.)   In long grass, under bark, under logs and rocks, at edges of fields, at lights, in trees and in damp basements.

Isoptera (termites)   At the base of fence posts, under and in rotting wood, under dung in pastures, in weed stems and trees, and in discarded boxes and newspaper.

Psocoptera (psocids and booklice)   In old books or papers, on tree trunks.

Anoplura (sucking lice)   On livestock and rodents.

Mallophaga (chewing lice)   Generally on birds.

Hemiptera (true bugs)   In grass and weed patches, under rocks, near or in water, under bark and leaves, at lights, and in gardens.

Homoptera (aphids, leafhoppers, etc.)   Around shrubbery, grass, weed patches, and many crops.

Thysanoptera (thrips)   Common in most flowers.

Coleoptera (beetles)   In trees and shrubbery, under rocks, beneath bark, in rotting logs, under decaying matter, in water, at lights, around flowers.

Neuroptera (lacewings, dobsonflies, etc.)   Near or in water, on trees or vegetation, at lights.

Mecoptera (scorpionflies)   Near streams, in dense woods and vegetation.

Lepidoptera (butterflies, skippers, and moths)   At flowers, on trees, and on many types of vegetation as larvae.

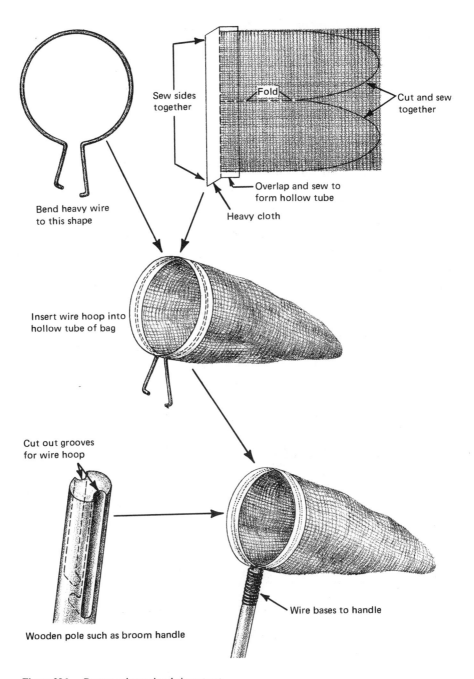

Sew sides
together

Fold

Cut and sew
together

Overlap and sew to
form hollow tube

Heavy cloth

Bend heavy wire
to this shape

Insert wire hoop into
hollow tube of bag

Cut out grooves
for wire hoop

Wire bases to handle

Wooden pole such as broom handle

Figure 296. Constructing a simple insect net.

Diptera (flies, etc.)   At lights, on flowers, near water, about livestock and fecal material, near decaying matter.

Hymenoptera (bees, wasps, ants, etc.)   On flowers, near mud banks, at lights, under stones and wood, on shrubbery and trees, around barns and houses, under bridges.

Kill Bottle

A suggested design of a kill bottle is provided in Figure 297. Any size bottle can be used, but the most useful sizes are half-pint and pint wide-mouth jars. Thick-walled bottles are best because they are less likely to break. After the plaster of Paris has been poured, allow the mixture to dry for several days with the lid completely off. Fumigant, such as carbon tetrachloride or ethyl acetate, may be added after the plaster is dry. A layer of fumigant 1 cm deep will be completely absorbed in a standard pint kill bottle. DO NOT INHALE

Tight screw-lid or cork. (Do not use a rubber stopper which may swell and break the bottle.)

Label is very important.

Few loose strips of crumpled paper. These absorb water and help prevent damage to the insects. Change paper when it becomes moist.

Layer of plaster of Paris or several layers of blotter paper tightly pressed down. Plaster layer should be 0.6 to 1.3 cm thick. Add plaster to water, stir rapidly; when as thick as heavy cream pour into bottle. Before plaster has set completely poke several small holes in it with a wire or toothpick. These allow the carbon tetrachloride to flow to the bottom of the jar more quickly.

Layer of sawdust 0.6 to 1.3 cm thick. This absorbs the chemical used for killing and provides room for the rubber to expand.

Rubber bands or other pieces of rubber cut into pieces not more than one inch in length. Part of an old inner tube is an excellent source of rubber. This layer also should be 0.6 to 1.3 cm thick.

Tape wrapped around bottom will help prevent breakage. Adhesive tape, electrical tape, or masking tape are all suitable.

Figure 297.   One way of constructing a kill bottle.

THE FUMIGANT. LABEL THE BOTTLE POISON. A few strips of blotting paper placed in the completed bottle will absorb excess moisture and keep the interior dry.

As soon as hard-bodied insects are collected, they should be placed in the kill bottle. Allow them to remain for approximately one hour to insure death. During a collecting trip in which many kinds and numbers of insects are being collected, it is wise to have several bottles or to remove the more delicate insects to prevent damage by larger insects. Large butterflies, moths, beetles, and dragonflies may be placed in a separate bottle with a piece of cotton saturated with fumigant or they may be injected with alcohol or formalin by using a syringe.

Soft-bodied insects, especially immature stages, should not be placed in the kill bottle but killed by inserting them into or injecting them with alcohol. Although ethyl or isopropyl alcohols are adequate for preserving these fragile insects for future observations, preservatives are available which are preferred if the specimen is to be placed in a research collection.

Relaxing

If, for any reason, collected insects cannot be mounted until they are dry, relaxing them will prevent breakage. A relaxing jar is easily constructed by placing 5 cm of sand in the bottom of a glass jar. The sand is then saturated with water, to which a few drops of formalin or carbolic acid crystals have been added to prevent mold growth, and a piece of blotting paper is placed on the sand surface to receive the specimens. The lid is then closed for from one to three days, depending on the size of the insects.

Pinning

Much of the appearance of the insect and ease in viewing diagnostic characters depends on the neatness and care with which pinning is performed. Figure 298 shows where the pin is to be inserted in some of the major orders. Pin all Coleoptera through the right elytron, one quarter the distance from the base of the wing. Hemiptera and Homoptera should be pinned through the scutellum. Other orders should be pinned through the mesothorax. Specimens that are too small and delicate to be pinned normally are mounted either on cardboard points or on micropins. If points are used, the insect is touched by the point to which a small dab of shellac, nail polish, or glue has been added and positioned to the left of the insect pin with the head forward (Fig. 298). Micropins are short, very delicate headless pins that should pierce the insect from the ventral side. Care should be taken so that the pin does not penetrate the dorsum. The other side of the micropin is in turn connected to the insect pin through bits of cork or pith.

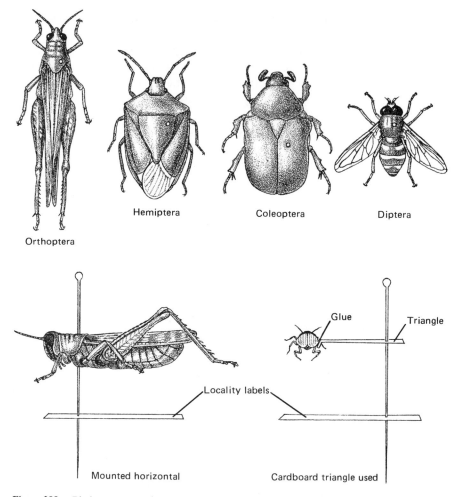

Figure 298.   Pin insects correctly.

It is best to spread the wings of moths and butterflies. First, pin the insect through the mesothorax and insert the pin in the crease of the pinning board (Fig. 299). Move the forewing anteriorly until the back margin forms a right angle with a line parallel with the long axis of the body. Hold the wing in position temporarily with a pin behind the large anterior veins. Move the hindwings forward to their proper position. Next, cut small strips of paper and place them across the wings and pin around the wing outline (exercise care not to penetrate the wings). The temporary pins may now be removed. Two or three days should be sufficient, except for large individuals, to permit thorough drying in dry climates.

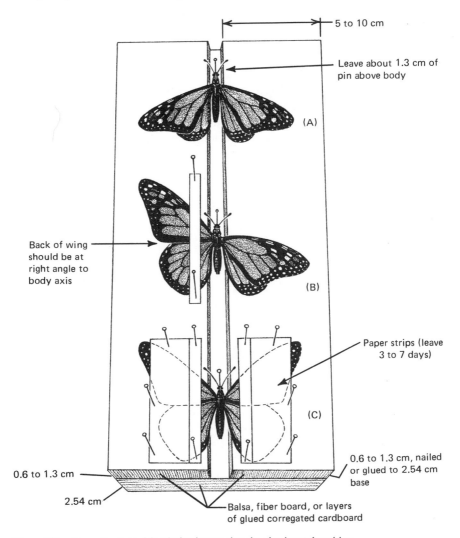

5 to 10 cm

Leave about 1.3 cm of
pin above body

(A)

Back of wing
should be at
right angle to
body axis

(B)

Paper strips (leave
3 to 7 days)

(C)

0.6 to 1.3 cm, nailed
or glued to 2.54 cm
base

0.6 to 1.3 cm

2.54 cm

Balsa, fiber board, or layers
of glued corregated cardboard

Figure 299.   Spreading board for drying insect wings in a horizontal position.

Labeling

All specimens should bear a label giving the locality and date of capture
(Fig. 298). Be accurate, for incorrect information is worse than no data. If you
do not know the exact date, include only that which is definite. Dates should
be written in one of the following styles: 10 Aug. 1977, 10.VIII.1977, or
VIII.10.1977. Never use 8/10/77 or 10/8/77. A suggested form is:

KANSAS: Riley Co.
Manhattan
VIII.10.1977

     Labels placed on insect pins should not exceed 18 mm by 5 mm in size. Names of collectors are placed on a second label beneath the locality label. Photographic collecting labels may be secured by filling a typewritten sheet with columns of the various data desired. This sheet is then reduced by photography and printed. Computers can be programmed to produce the necessary labels.

Glossary

Abdomen: The major region of the body posterior to the thorax or cephalothorax.

Adaptability: Capacity for evolutionary change.

Adaptive radiation: Evolution and spread of a population of organisms into different ecological niches.

Aerobic respiration: The oxidative process of releasing energy within the cell.

Aeropyle: Specialized pores on the egg surface for oxygen uptake.

Aggregation: Grouping by a species resulting from some attractive stimulus.

Allele: The alternate forms of a gene, having the same locus in homologous chromosome.

Alimentary canal: The tube between the mouth and anus in which food movement, digestion, and absorption occurs.

Amino acid: Any organic acid containing an amino radical (NH_2); the building stones of proteins.

Anaerobic respiration: Releasing energy within the cell without oxygen.

Anamorphic development: Development by adding segments to the body.

Antenna (pl. antennae):   A pair of segmented sensory appendages on the head.

Anus: The posterior opening of the alimentary canal.

Arista: Large bristle on the third segment of the antennae of specialized flies.

Arthropoda: A phylum of animals that have segmented body and legs and an exoskeleton.

Asexual reproduction: Not involving the fusion of nuclei of different gametes.

ATP: Adenosine triphosphate; a molecule having high-energy bonding through a phosphate group.

Bilateral symmetry: Symmetry in which the body can be divided by a median plane into mirror images of one another.

Biomass: The amount of living matter.

Biome: Major communities of animals and plants having a similar life form.

Blastoderm: The cells produced through cleavage of the zygote.

Blastokinesis: Movement or migration of the embryo within the egg.

Carnivore: An eater of animals.

Cephalothorax: A body region having the characteristics of both a head and thorax.

Cercus (pl. cerci): Appendage of the eleventh abdominal segment.

Chelicera (pl. chelicerae): Dominant feeding appendages of the third body segment.

Chitin: A nitrogenous polysaccharide with an empirical formula of $(C_8H_{13}O_5N)$.

Circadian rhythm: Physiological rhythm with an approximate 24-hour cycle.

Clypeus: Sclerite beneath the frons.

Coarctate pupa: Pupa remains within the last larval exoskeleton, the puparium.

Community: All plants and animals living in a specific region.

Complete metamorphosis: Holometabolic development; immatures (larvae) do not closely resemble the adults and a pupal stage is present.

Compound eye: Optic organ formed from ommatidia.

Coxa: Basal segment of a leg.

Cryptic coloration: Coloration of organism that blends with background.

Cuticle: The secreted part of the integument.

Density-dependent: Factor whose influence varies with the density of the population.

Density-independent: Factor whose influence does not vary with the number of individuals in a population.

Diapause: A state of hibernation or suspended development in insects.

Digestion: Enzymatic hydrolysis of food into absorbable size molecules.

Dioecious: Condition in which male and female reproductive organs are in separate individuals.

Diploid: Possessing a double set of chromosomes (2n).

Diurnal: Active during daytime hours.

Diversity: Measure of variety of species that takes into account the relative abundance of each species.

Ecdysial suture: Weakened area of integument that splits during molting to permit the old exoskeleton to be shed.

Ecdysis: The process of shedding an exoskeleton.

Eclosion: Process of an individual's emerging or hatching from an egg.

Ecological efficiency: Percent of energy in biomass produced by a trophic level that is incorporated into biomass by the next higher trophic level.

Ecological niche: The way of life of an organism including food, habitat requirements, and so forth.

Ecosystem: The living and nonliving factors and their interactions in an area.

Ectoderm: The outer germ layer of an early embryo.

Elytron (pl. elytra):   Thickened and hardened front wings of beetles.

Endocrine: A ductless gland producing hormones.

Endoderm: The innermost layer of cells produced in an early embryo that becomes part of the lining of the alimentary canal.

Environment: The total surroundings of an organism.

Enzyme: An organic catalyst.

Epimorphic development: Immatures possess same number of body segments as adults.

Estivation: Reduction of biological activity by an organism in response to periods of hot or dry weather.

Exarate pupa: Pupa in which the appendages are not appressed to the body.

Exoskeleton: A skeleton on the outside of the body.

Femur (pl. femora):   Third segment of the leg from the body.

Food chain: The direction of energy flow between species that feed upon one another.

Frons: Area between eyes and bordered above by the vertex and below by the clypeus.

Galea: Outer distal lobe of the maxilla.

Gamete: A mature reproductive cell, either sperm or ovum.

Ganglion (pl. ganglia):   A concentration of motor nerve cell bodies and association neurons and performing nervous coordination.

Gena (pl. genae):   Cheek area of insect head below compound eye.

Genetic drift: Change in allele frequency as a result of random occurrences, variations in fecundity, and mortality in a population.

Gene pool: The entire genetic complement.

Haploid: Possessing a single set of chromosomes (n).

Hemelytron (pl. hemelytra):   Front pair of wings of true bugs (Hemiptera); the basal part is thickened and the distal part is membranous.

Hemocoel: Blood cavity not entirely lined by mesoderm.

Herbivore: An organism that consumes living plant material.

Heterozygous: Containing two forms (alleles) of a gene, one from each parent.

Homozygous: Containing identical alleles of a gene.

Hyperosmotic: Having a salt concentration greater than that of the surrounding medium.

Hypopharynx: "Tongue-like" structure anterior to labium.

Hyposmotic: Having a salt concentration less than that of the surrounding medium.

Imaginal disc: Clusters of undifferentiated cells in larvae that will give rise to adult structures.

Incomplete metamorphosis: Hemimetabolic development; the immatures (nymphs) do not resemble the adults and a pupal stage is lacking.

Instar: The insect from one molt to the next.

Integument: The body wall or exoskeleton.

Intima: The extension of the exoskeleton into the internal cavities of the body for lining the foregut, hindgut, and tracheae.

Irritability: The ability to respond to stimuli.

Kin selection: Selection operating between closely related individuals to produce cooperation.

K-selection: Favoring population densities near carrying capacity (K).

Labellum (pl. labella):   Modified tip of the labium of some flies.

Labium (pl. labia):   Fused third pair of appendages in insect mouthparts.

Labrum: Sclerite beneath the clypeus; serves as upper lip in insects that have chewing mouthparts.

Lacinia: Inner distal lobe of maxilla.

Larva: Immature stage in life cycle; often restricted to specific forms such as feeding immatures in insects that have complete metamorphosis.

Life cycle: The sequence of events from egg to adult.

Mandibles: Appendages of the fourth body segment that become the first pair of mouthparts in mandibulate species.

Maxilla (pl. maxillae):   Second pair of feeding appendages in mandibulate species.

Meiosis: Process of reducing the number of chromosomes from diploid to haploid in gamete formation.

Mesenteron: The middle portion of the alimentary canal that forms as a blind sac.

Mesoderm: The embryonic group of cells forming between the ectoderm and endoderm.

Mesothorax: The second or middle segment in the insect thorax.

Metamere: A somite or segment.

Metamorphosis: Change in shape from one stage to another in life cycle.

Metathorax: Third segment in insect thorax.

Mimicry: Superficial resemblances between two organisms that result in protection for at least one organism, the mimic.

Molting: The process of shedding an exoskeleton in arthropods.

Mutualism: Symbiotic relationship requiring both organisms for survival.

Mutation: Process that results in gene changes from one heritable allele to another heritable allele.

Natural selection: Differential survival and reproduction of individuals.

Neoptera: Insects capable of flexing or folding their wings.

Neuron: A nerve cell; one highly specialized for conducting cellular impulses.

Nocturnal: Pertaining to the night hours.

Notum (pl. nota):   The dorsal plate or tergum of each thoracic segment.

Nymph: The immature stage in incomplete metamorphosis.

Obtect pupa: Pupa that has appendages closely appressed to the body.

Occiput: Portion of head behind the vertex, eyes, and gena.

Ocellus (pl. ocelli):   Photoreceptor located between compound eyes in adults and nymphs.

Ommatidium (pl. ommatidia):   Visual unit of a compound eye.

Omnivore: An organism that has a broad diet including both plant and animal material.

Osmoregulation: Regulation of the salt and water concentration in cells and organisms.

Oviparous: Egg-laying with hatching outside the body of the female.

Oviposition: Placing of eggs in suitable position and habitat.

Ovipositor: Egg-laying organ.

Ovoviviparous: Eggs hatch within the female's body.

Paleoptera: Insects that are not capable of flexing or folding their wings.

Palpus (pl. palpi): Segmented appendage on maxillae and labium.

Parasite: An organism living at the expense of another.

Paranotum (pl. paranota): Notal flanges in fossil insects.

Parthenogenesis: Development from an unfertilized egg.

Pedipalp (pl. pedipalpi): Second pair of appendages in Arachnida; homologous to mandibles.

pH: Symbol denoting relative concentration of hydrogen ions in solution; a scale of relative acidity and alkalinity.

Pheromones: Chemical substances used for communication between individuals.

Photoperiodism: Seasonal responses to changes in length of daylight.

Phylum (pl. phyla): A category of classification comprised of closely related classes.

Plastron: Structure to obtain dissolved oxygen.

Pleural sulcus: Groove running dorsally from coxa; often referred to as pleural suture.

Pleuron: Lateral plate in an insect segment.

Pollination: Transfer of male gametophyte (pollen) to female flower or female part of flower.

Population: All organisms sharing a gene pool.

Predator: Organism that feeds on another.

Proctodeum: Posterior division of alimentary canal formed from invagination.

Prothorax: First segment of insect thorax.

Ptilinum: Eversible sac in some Dipterous adults that is used to rupture the puparium.

Pupa: Intermediate stage between larva and adult in complete metamorphosis.

Puparium: Covering of pupa in some Dipterous species; represents last larval exoskeleton.

r-selection: Favoring rapid population growth at low density.

Sclerite: Plate in body wall formed from either sclerotization or calcification.

Sclerotization: Process of hardening integument by the tanning of the protein sclerotin.

Search image: Behavior that enables predators to increase ability to locate prey.

Seta (pl. setae): Hair-like processes from integument.

Sibling: One of two or more individuals having common parents.

Solitary: Living alone.

Speciation: Separation of one population into two or more reproductively isolated independent evolutionary units.

Spermatophore: Gelatinous covered packets containing sperm.

Spermiogenesis: Process of transforming spermatids into functional sperm cells.

Spiracle: Openings into tracheal system.

Stemmata: Photoreceptors in larval insects and some Ametabolous adults; often referred to as lateral ocelli.

Stomodeum: Anterior portion of alimentary canal formed from invagination.

Stylet: Mouthpart modified into a pointed or dagger-like structure.

Sulcus (pl. sulci): Linear braces, sometimes referred to as sutures.

Surface/volume ratio: A ratio permitting comparisons to be made on effects of size.

Symbiosis: Close association between two organisms.

Tagmosis: Specialization of body into regions.

Tarsus (pl. tarsi): Last or distal part of an insect leg.

Tegmen (pl. tegmina): Parchment-like textured forewings of Orthoptera.

Tergum (pl. terga): Dorsal part of a body segment; often referred to as the notum in thoracic segments.

Tibia (pl. tibiae): Fourth segment of an insect leg.

Trachea (pl. tracheae): Tubes used to transport air throughout the body.

Tracheole: The untapering terminal tracheae in which gaseous exchange occurs.

Trochanter: Second leg segment from the body.

Tympanum (pl. tympana): Membrane that serves as an eardrum or covers the auditory organs.

Urticating setae: Setae capable of producing irritation in another organism.

Vertex: Top of the insect head.

Voltinism: The number of generations per year.

Selected References

ADAMS, LOWELL, 1970. *Population Ecology.* Belmont, Calif.: Dickenson, 160 pp.

AHEARN, GREGORY, 1970. Control of Water Loss in Beetles. *J. Exp. Biol., 53:*573–95.

ALEXANDER, R. D., 1974. The Evolution of Social Behavior. *Ann. Rev. Syst. Ecol., 4:*325–83.

ALVARADO, CARLOS A., and L. J. BRUCE-CHWATT, 1962. Malaria. *Scientific Amer., 206(5):*86–98.

ANDREWARTHA, H. G., 1971. *Introduction to the Study of Animal Population,* 2nd ed. Chicago: Univ. Chicago Press, 275 pp.

BAKER, HERBERT G., 1968. Intrafloral Ecology. *Ann. Rev. Entomol., 13:*385–414.

BARRINGTON, E. J. W., 1967. *Invertebrate Structure and Function.* London: Thomas Nelson, 549 pp.

BATRA, SUZANNE W. T., and LEKH R. BATRA, 1967. The Fungus Gardens of Insects. *Scientific Amer., 217(5):*112–20.

BEAMENT, J. W. L., and J. E. TREHERNE, 1967. *Insects and Physiology.* New York: American Elsevier, 364 pp.

BECK, STANLEY D., 1968. *Insect Photoperiodism.* New York: Academic Press, 282 pp.

BEIRNE, BRYAN P., 1963. *Collecting and Preserving Insects.* Ottawa: Canada Dept. Agric. Publ. 932.

BEROZA, MARTIN, 1970. *Chemicals Controlling Insect Behavior.* New York: Academic Press, 170 pp.

BISHOP, J. A., and LAURENCE M. COOK, 1975. Moths, Melanism and Clean Air. *Scientific Amer., 232(1):*90–99.

BLEST, A. D., 1957. The Evolution of Protective Displays in the Saturnoidea and Sphihgidae (Lepidoptera). *Behavior,* 11:257–309.

———, 1963. Longevity, Palatability, and Natural Selection in Five Species of New World Saturnid Moths. *Nature,* 197:1183–86.

BODENHEIMER, F., 1951. *Insects as Human Food.* The Hague: W. Junk, 353 pp.

BOER, P. J., and G. R. GRADWELL, 1970. *Dynamics of Populations: Proceedings of the Advanced Study Institute on "Dynamics of Numbers in Populations,"* Oosterbeck, Netherlands; Wageningen: Centre for Agri. Publ. & Documentation.

BORROR, DONALD J. and DWIGHT M. DELONG, 1971. *An Introduction to the Study of Insects,* 3rd ed. New York: Holt, Rinehart and Winston, 812 pp.

BORROR, DONALD J., DWIGHT M. DELONG, and CHARLES A. TRIPLEHORN, 1976. *An Introduction to the Study of Insects,* 4th ed. New York: Holt, Rinehart and Winston, 852 pp.

BORROR, DONALD J., and RICHARD E. WHITE, 1970. *A Field Guide to the Insects of America North of Mexico.* Boston: Houghton Mifflin, 404 pp.

BOUGHEY, A. S., 1971. *Fundamental Ecology.* San Francisco: Intext Educational Publ., 222 pp.

BOWERS, W. S., T. OHTA, J. S. CLEERE, and P. A. MARSELLA, 1976. Discovery of Insect Ant-Juvenile Hormones in Plants. *Science,* 193:542–47.

BROWER, L. P., 1969. Ecological Chemistry. *Scientific Amer.,* 220(2):22–29.

BROWER, L. P., J. V. Z. BROWER, and C. T. COLLINS, 1963. Relative Palatability and Mullerian Mimicry Among Neotropical Butterflies of the Subfamily Helioconiinae. *Zoologica,* 48:65–84.

BROWN, A. W. A., 1971. *Insecticide Resistance in Arthropods.* World Health Organization Monograph Series No. 38.

BROWN, FRANK A., 1972. The "Clocks" Timing Biological Rhythms. *Amer. Scientist,* 60(6): 756–66.

BRUES, CHARLES T., 1972. *Insects, Food, and Ecology.* New York: Dover, 465 pp.

BUCK, JOHN, 1962. Some Physical Aspects of Insect Respiration. *Ann. Rev. Entomol.,* 7:27–56.

BURSELL, E., 1970. *An Introduction to Insect Physiology.* New York: Academic Press, 276 pp.

BUSHLAND, R. C., 1975. Screwworm Research and Eradication. *Bul. Entomol. Soc. Amer.,* 21(1):23–26.

BUTLER, C. G., 1967. Insect Pheromones. *Biol. Rev.,* 42:42–87.

CANTWELL, G. E., ed., 1974. *Insect Diseases,* Vols. I and II. New York: Marcell Decker, 595 pp.

CARSON, RACHEL, 1962. *Silent Spring.* Greenwich, Conn. Fawcett, 304 pp.

CARTHY, J. D., 1956. *The Behavior of Arthropods.* San Francisco: W. H. Freeman, 148 pp.

CHAPMAN, R. F., 1969. *The Insects Structure and Function.* New York: American Elsevier, 819 pp.

CHAUVIN, REMY, 1967. *The World of an Insect.* New York: McGraw-Hill, 256 pp.

COPE, OLIVER B., 1971. Interactions Between Pesticides and Wildlife. *Ann. Rev. Entomol.,* 16:325–64.

CUMMINGS, K. W., 1973. Trophic Relations of Aquatic Insects. *Ann. Rev. Entomol.,* 18:183–206.

DAUMER, K., 1958. Blumenfarben wie sie die Bienen sehen. *Zeitschrift fur vergleichende Physiologie,* 41:49–110.

DECK, ERRETT, 1975. Federal and State Pesticide Regulations and Legislation. *Ann. Rev. Entomol.,* 20:119–31.

DECKER, GEORGE C., 1960. Insecticides in the 20th-Century Environment. *A.I.B.S. Bull.,* 10(2):27–31.

DETHIER, V. G., 1963. *The Physiology of Insect Senses.* New York: Wiley, 266 pp.

———, 1971. A Surfeit of Stimuli: A Paucity of Receptors. *Amer. Scientist,* 59(6):706–15.

DOWNES, J. A., 1965. Adaptations of Insects in the Arctic. *Ann. Rev. Entomol.,* 10:257–74.

EBERHARD, MARY JANE WEST, 1975. The Evolution of Social Behavior by Kin Selection. *Quarterly Rev. Biol.,* 50(1):1–33.

EDMUNDS, G. F., AND J. R. TRAVER, 1954. The Flight Mechanics and Evolution of the Wings of Ephomoroptera, with Notes on the Archeotype Insect Wing. *Jour. Wash. Acad. Science,* 44:390–99.

EDNEY, E. B., 1957. *The Water Relations of Terrestrial Arthropods.* Cambridge: Cambridge Univ. Press, 105 pp.

EDWARDS, CLIVE, 1969, Soil Pollutants and Soil Animals. *Scientific Amer.,* 220(4):88–99.

EHRLICH, PAUL R., and PETER H. RAVEN, 1963. Butterflies and Plants: A Study in Coevolution. *Evolution,* 18(4):586–608.

ENGLEMANN, FRANZ, 1970. *The Physiology of Insect Reproduction.* New York: Pergamon, 307 pp.

EVANS, HOWARD E., and MARY JANE WEST EBERHARD, 1970. *The Wasps.* Ann Arbor: Univ. Mich. Press, 265 pp.

FARB, PETER, and the editors of *Life,* 1962. *The Insects.* New York: Time, Inc., 192 pp.

FERNALD, H. T., and H. H. SHEPHERD, 1942. *Applied Entomology,* 4th ed. New York: McGraw-Hill, 400 pp.

FRAENKEL, G. S., and D. L. GUNN, 1940. *The Orientation of Animals.* Oxford: Clarendon, 352 pp.

FRAZER, J. F. D., and M. ROTHSCHILD, 1960. Defense Mechanisms in Warningly-coloured Moths and Other Insects. XI International Kongress für Entomologie, Wien: *Symposium,* 4:249–56.

FRAZIER, CLAUDE A., 1969. *Insect Allergy: Allergic and Toxic Reactions to Insects and Other Arthropods.* St. Louis, MO.: Warren H. Green, 493 pp.

FREE, JOHN B., *Insect Pollinators of Crops.* New York: Academic, 544 pp.

FRIDOVICH, IRWIN, 1975. Oxygen: Boon or Bane. *Amer. Scientist,* 63(1):54–59.

FRISCH, KARL VON, 1962. Dialects in the Language of the Bees. *Scientific Amer.,* 207(2):78–87.

———, 1971. *Bees—Their Vision, Chemical Senses, and Language.* Ithaca: Cornell Univ. Press, 157 pp.

FROST, S. W., 1959. *Insect Life and Insect Natural History,* 2nd ed. New York: Dover, 526 pp.

GIBBS, A. J., 1974. *Viruses and Invertebrates.* New York: American Elsevier, 673 pp.

GILLET, J. D., 1973. The Mosquito: Still Man's Worst Enemy. *Amer. Scientist,* 61(4):430–36.

GOSS, RICHARD J., 1969. *Principles of Regeneration.* New York: Academic, 287 pp.

HADLEY, NEIL F., 1972. Desert Species and Adaptations. *Amer. Scientist,* 60(3):338–47.

HAGAN, HAROLD R., 1951. *Embryology of the Viviparous Insects.* New York: Ronald, 472 pp.

HAMILTON, K. G. ANDREW, 1971. The Insect Wing, Part I. Origin and Development from Notal Lobes. *Kans. Entomol. Soc.,* 44(4):421–33.

HAMILTON, W. D., 1964. The Genetic Theory of Social Behavior, 1. *J. Theoret. Biol.,* 7:1–16.

———, 1964. The Genetic Theory of Social Behavior, 2. *J. Theoret. Biol.,* 7:17–52.

HAWKINS, FRANK, 1970. The Clock of the Malaria Parasite. *Scientific Amer.,* 222(6):123–31.

HICKEM, E. NORMAN, 1973. *Insect Factory in Wood Decay.* New York: St. Martins.

HINTON, H. E., 1963. The Origin and Function of the Pupal State. *Proc. Royal Entom. Soc. London,* 38:77–85.

HINTON, H. E., 1969. Respiratory Systems of Insect Egg Shells. *Ann. Rev. Entomol.,* 14:343–68.

HOFFMANN, C. H., 1974. Critical Issues Face Entomologists. *Bul. Entomol. Soc. Amer.,* 20(4):316:18.

HOLLING, C. S., 1965. The Functional Response of Predators to Prey Density and Its Role in Mimicry and Population Regulation. *Mem. Entomol. Soc. Canada,* 45:1–60.

HUTCHINS, ROSS, 1966. *Insects.* Englewood Cliffs: Prentice-Hall, 324 pp.

HYNES, H. B. N., 1970. The Ecology of Stream Insects. *Ann. Rev. Entomol.,* 15:25–42.

IMMS, A. D., 1957. *A Textbook of Entomology,* 9th ed. London: Methuen, 886 pp.

JACOBSON, MARTIN, 1972. *Insect Sex Pheromones.* New York: Academic, 382 pp.

JACOBSON, MARTIN, and MORTON BEROZA, 1964. Insect Attractants. *Scientific Amer.,* 211(2):20–27.

JAMES, MAURICE T., and ROBERT F. HARWOOD, 1969. *Herm's Medical Entomology,* 6th ed. London: Macmillan, 484 pp.

JANZEN, D. H., 1970. Herbivores and the Number of Tree Species in Tropical Forests. *Amer. Nat.,* 104:501–28.

JOHANNSEN, OSKAR A., and FERDINAND H. BUTT, 1941. *Embryology of Insects and Myriapods.* New York: McGraw-Hill, 462 pp.

JOHNSON, C. G., 1963. The Aerial Migration of Insects. *Scientific Amer.,* 209(6):132–38.

———, 1966. A Functional System of Adaptive Flight. *Ann. Rev. Entomol.,* 11:233–53.

JUKES, THOMAS H., 1963. People and pesticides. *Amer. Scientist,* 51(3):335–61.

KETTLEWELL, H. B. D., 1959. Darwin's Missing Evidence. *Scientific Amer.,* 200(3):48–53.

KLOTTS, A. B., and E. B. KLOTTS, 1965. *Living Insects of the World.* London: Hamish Hamilton, 304 pp.

KNIPLING, E. F., 1959. Sterile-male Method of Population Control. *Science,* 130:902–04.

KONISHI, MASAKAZU, 1971. Ethology and Neurobiology. *Amer. Scientist,* 59(1):56–63.

KRISHNA, K., and F. M. WEESNER, 1969. *Biology of Termites,* Vols. 1 and 2. New York: Academic, 598 and 643 pp.

LEE, K. E., and T. G. WOOD, 1971. *Termites and Soils.* New York: Academic, 251 pp.

LÜSCHER, MARTIN, 1961. Air-Conditioned Termite Nests. *Scientific Amer.,* 205(1):138–45.

_____, 1961. Social Control of Polymorphism in Termites. In *Insect Polymorphism,* pp. 57–67, Symposium no. 1, R. Ent. Soc. Lond.

MANN, JOHN, 1969. Cactus-Feeding Insects and Mites. *Smithsonian Institute Bul. 256.*

MATSUDA, RYUICHI, 1965. *Morphology and Evolution of the Insect Head.* Ann Arbor: *Memoirs of the Amer. Entom. Instit. No. 4,* 334 pp.

MCATEE, W. L., 1932. Effectiveness in Nature of the So-called Protective Adaptations in the Animal Kingdom, Chiefly as Illustrated by the Food Habits of Nearctic Birds. *Smithsonian Misc. Coll.* 85(7):201 pp.

MENAKER, MICHAEL, 1969. Biological Clocks. *Bioscience,* 19(8)681–92.

METCALF, C. L., W. P. FLINT, and R. L. METCALF. *Destructive and Useful Insects, Their Habits and Control.* New York: McGraw-Hill, 1071 pp.

MIALL, L. C., 1903. *The Natural History of Aquatic Insects.* New York: Macmillan, 395 pp.

MICHENER, CHARLES D., 1958. The Evolution of Social Behavior in Bees. *Proc. Tenth Int. Congr. Entomol., Montreal,* 2:441–47.

_____, 1969. Social Behavior of Bees. *Ann. Rev. Entomol.,* 14:299–342.

MICHENER, C. D., and M. H. MICHENER, 1951. *American Social Insects.* New York: Van Nostrand, 267 pp.

NEWSOM, L. D., 1967. Consequences of Insecticide Use on Nontarget Organisms. *Ann. Rev. Entomol.,* 12:323–46.

NUESCH, HANS, 1968. The Role of the Nervous System in Insect Morphogenesis and Regeneration. *Ann. Rev. Entomol.,* 13:27–44.

ODUM, E. P., 1971. *Fundamentals of Ecology,* 3rd ed. Philadelphia: Saunders, 575 pp.

OSSIANNILSSON, FREJ, 1966. Insects in the Epidemiology of Plant Viruses. *Ann. Rev. Entomol.,* 11:213–32.

PAINTER, R. H., 1968. *Insect Resistance in Crop Plants.* Lawrence: Univ. Press of Kansas, 544 pp.

PAMPANA, E. J., and P. F. RUSSELL, 1955. *Malaria, a World Problem.* Geneva: World Health Organization, 72 pp.

PARRISH, H. M., 1963. Analysis of 460 Fatalities from Venomous Animals in the U.S. *Jour. Med. Science,* 245(2):129–41.

PEDIGO, LARRY, ed., 1972. *Insect Ecology and Population Management: Readings in Theory, Technique, and Strategy.* New York: Mss Educational Publ. Co., 309 pp.

PETERSON, ALVAH, 1960. *Larvae of Insects,* Part II. *Coleoptera, Diptera, Neuroptera, Siphonaptera, Mecoptera, Trichoptera.* Columbus, Ohio: privately published by author, 416 pp.

_____, 1962. *Larvae of Insects.* Part I. *Lepidoptera and Plant-Infesting Lepidoptera and Hymenoptera.* Columbus, Ohio: privately published by author, 315 pp.

PHILIP, CORNELIUS B., and WILLY BURGDORFER, 1961. Arthropod Vectors as Reservoirs of Microbial Disease Agents. *Ann. Rev. Entomol.,* 6:391–412.

PIANKA, E. R., 1974. *Evolutionary Ecology.* New York: Harper & Row, 356 pp.

PINHEY, E. C. G., 1968. *Introduction to Insect Study in Africa.* London: Oxford Univ. Press, 235 pp.

POMERANTZ, C., 1959. Arthropods and Psychic Disturbances. *Bull. Entomol. Soc. Amer.,* 5:65–67.

POWELL, J. H., 1949. *Bring Out Your Dead. The Great Plague of Yellow Fever in Philadelphia in 1793.* Philadelphia: Univ. Penn. Press, 304 pp.

RATHCKE, BEVERLY J., and ROBERT W. POOLE, 1974. Coevolutionary Race Continues: Butterfly Larval Adaptation to Plant Trichomes. *Sciences,* 187:175–76.

REED, CHESTER, 1943. *Land Birds East of the Rockies.* Garden City: Doubleday.

REMINGTON, C. L., 1963. Historical Background of Mimicry. *Proc. XVI Int. Congr. Zool. Wash.,* 4:145–49.

RENTZ, DAVID C., 1972. The Lock and Key as an Isolating Mechanism in Katydids. *Amer. Scientist,* 60(6):750–55.

RETTENMEYER, CARL W., 1963. Behavioral Studies of Army Ants. *Univ. Kans. Science Bul.,* 44(9):281–465.

_____, 1970. Insect Mimicry. *Ann. Rev. Entomol.,* 15:43–74.

RICKLEFS, ROBERT E., 1973. *Ecology.* Portland: Chiron Press, 861 pp.

ROEDER, KENNETH D., 1970. Episodes in Insect Brains. *Amer. Scientist,* 8:378–89.

SALT, R. W., 1961. Principles of Insect Cold-Hardiness. *Ann. Rev. Entomol.,* 6:55–74.

SCHALLER, FRIEDRICH, 1968. *Soil Animals.* Ann Arbor: Univ. Mich. Press. 144 pp.

_____, 1971. Indirect Sperm Transfer by Soil Arthropods. *Ann. Rev. Entomol.,* 16:406–46.

SCHOONHOVEN, L. M., 1968. Chemosensory Bases of Host Plant Selection. *Ann. Rev. Entomol.,* 13:115–36.

SEXTON, O. F., 1960. Experimental Studies of Artificial Batesian Mimics. *Behavior,* 15:244–52.

SMITH, DAVID S., 1968. *Insect Cells, Their Structure and Function.* Edinburgh: Oliver & Boyd, 372 pp.

SMITH, KENNETH M., 1958. Transmission of Plant Viruses by Arthropods. *Ann. Rev. Entomol.,* 3:469–82.

SNODGRASS, R. E., 1935. *Principles of Insect Morphology.* New York: McGraw-Hill, 667 pp.

_____, 1958. Evolution of Arthropod Mechanisms. *Smithsonian Misc. Coll.,* 138(2):1–77.

_____, 1965. *A Textbook of Arthropod Anatomy.* New York: Stechert-Hafner, 363 pp.

STAHL, G. B., 1975. Insect Growth Regulators with Juvenile Hormone Activity. *Ann. Rev. Entomol.,* 20:417–60.

SUDD, JOHN J., 1967. *An Introduction to the Behavior of Ants.* London: Arnold, 200 pp.

SWANS, L. A., 1964. *Beneficial Insects.* New York: Harper & Row, 429 pp.

TEAL, J. M., 1971. Community Metabolism in a Temperate Cold Spring. *Ecological Monog.,* 27:283–302.

THEILER, MAX, and W. G. DOWNS, 1973. *Arthropod Borne Viruses of Vertebrates.* New Haven, Conn.: Yale Univ. Press, 578 pp.

WALDRON, WILLIAM G., 1962. The Role of the Entomologist in Delusionary Parasitosis (Entomophobia). *Bull. Entomol. Soc. Amer.,* 8(2):81–83.

WALOFF, Z., 1966. *The Upsurges and Recessions of the Desert Locust Plague: An Historical Survey.* London: Anti-Locust Research Center, 111 pp.

WEAVER, NEVIN, 1966. Physiology of Caste Determination. *Ann. Rev. Entomol.* 11:79–102.

WELLS, MARTIN, 1968. *Lower Animals.* New York: McGraw-Hill, World Univ. Lib., 255 pp.

WHITTEN, M. J., and G. G. FOSTER, 1975. Genetic Methods of Pest Control. *Ann. Rev. Entomol.,* 20:461–76.

WICKLER, WOLFGANG, 1968. *Mimicry in Plants and Animals.* (Trans. by R. D. Martin). London: World University Library, Weidenfeld & Nicolson, 253 pp.

WIGGLESWORTH, V. B., 1954. *The Physiology of Insect Metamorphosis.* Cambridge: University Press, 149 pp.

_____, 1966. *The Life of Insects.* London: Weidenfeld & Nicholson, 359 pp.

_____, 1970. *Insect Hormones.* Edinburgh: Oliver & Boyd, 159 pp.

_____, 1972. *The Principles of Insect Physiology,* 7th rev. ed. New York: E. P. Dutton, 544 pp.

WILLIAMS, C. M., 1947. Physiology of Insect Diapause. II. Interaction Between the Pupal Brain and Prothoracic Glands in Metamorphosis of the Giant Silkworm, *Platysamia cecropia. Biol. Bull.,* 93(2):89–98.

_____, 1952. Physiology of Insect Diapause. IV. The Brain and Prothoracic Glands as an Endocrine System in the Cecropia Silkworm. *Biol. Bull.,* 103(1):120–38.

_____, 1958. The Juvenile Hormone. *Scientific Amer.,* 198(2):67–74.

WILSON, EDWARD O., 1963. Pheromones. *Scientific Amer.,* 208(5)100–14.

_____, 1963. The Social Biology of Ants. *Ann. Rev. Entomol.,* 8:345–68.

_____, 1971. *The Insect Societies.* Cambridge: Belknap Press of Harvard Univ. Press, 548 pp.

Index

Numbers in boldface type represent illustration-bearing pages. Many items in this index are defined in the glossary.

317